The
MELTING
WORLD

ALSO BY CHRISTOPHER WHITE

Skipjack: The Story of America's Last Sailing Oystermen

Chesapeake Bay: Nature of the Estuary

Endangered and Threatened Wildlife of the Chesapeake Bay Region

The
MELTING
WORLD

*A Journey Across America's
Vanishing Glaciers*

Christopher White

St. Martin's Press
New York

www.stmartins.com

Design by Phil Mazzone

Map by Paul J. Pugliese

Library of Congress Cataloging-in-Publication Data

White, Christopher P., 1956–
    The melting world : a journey across America's vanishing glaciers / Christopher White.—
First edition.
        pages   cm
    Includes bibliographical references.
    ISBN 978-0-312-54628-1 (hardcover)
    ISBN 978-1-250-02885-3 (e-book)
  1. Glaciers—Montana—Glacier National Park.   2. Glaciers—Rocky Mountains.   3. Global warming—Montana—Glacier National Park.   4. Global warming—Rocky Mountains. 5. White, Christopher P., 1956—Travel.   6. Fagre, Dan.   7. Climatologists—United States—Biography.   8. Ecologists—United States—Biography.   9. Glacier National Park (Mont.)—Environmental conditions.   10. Rocky Mountains—Environmental conditions.   I. Title.
    GB2425.M9W47 2013
    551.31'20978652—dc23

                                                                2013013453

St. Martin's Press books may be purchased for educational, business, or promotional use. For information on bulk purchases, please contact Macmillan Corporate and Premium Sales Department at 1-800-221-7945, extension 5442, or write specialmarkets@macmillan.com.

First Edition: September 2013

10   9   8   7   6   5   4   3   2   1

For my grandchildren—Samira, Wyatt, and Lilou—because
the world is always looking toward the morning

Contents

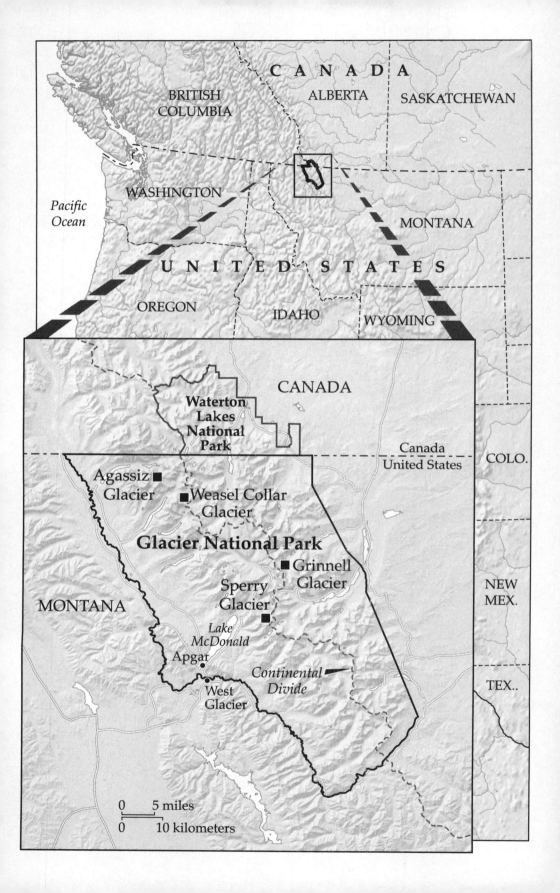

Author's Note

All of the people and places appearing in this book are real. However, for the sake of their privacy, I have adopted pseudonyms for some of the characters in play. One further exception: For simplicity, one of these players is a composite of three guides encountered on the trail. The sequence of a few events has been reshuffled slightly within the general time span of the story, 2008–2012. The USGS team depicted was actively engaged in field research for all those years, when the glacier survey was repeated annually.

Turning and turning in the widening gyre
The falcon cannot hear the falconer;
Things fall apart; the centre cannot hold;
Mere anarchy is loosed upon the world.

—from William Butler Yeats
The Second Coming (1920)

Gyre (pronounced *jire*)—n. A circle or spiral.
A swirling vortex. A circular current.

The
MELTING
WORLD

INTRODUCTION

The story of ice is the story of climate. Their histories walk in lockstep in mountain ranges around the world. Weather, which is climate's child, is what we see on any given day, wrapping around the peaks like a shawl. Sometimes hot, sometimes cold. On a warm day, melt water from ice feeds our rivers. Billions drink from it. During a cool summer, the majority of snow does not melt. It builds. From one snowflake, a glacier is born. Glaciers advance when winter outweighs each summer. They retreat under the opposite regime, as we have today. Through the ages, the parade of climate has been shaped by fluctuations and cycles, by flights of fancy as well as centuries-long freezes and thaws. The promenade of ice is climate's frozen twin. They give each other meaning.

Climate and ice also influence each other on a grand scale. For example, as the glaciers melt between ice ages or now in modern times, the sea levels rise, altering ocean currents, such as the Gulf Stream. The displaced currents, in turn, shift the climate. The loop is now complete. Scotland becomes warm. New England has an early spring, but that is only the beginning. Could Greenland melt? Could another hurricane, perhaps more fierce than Sandy, put New York City under siege again? As the climate warms, the story of destruction is repeated all over the Earth. The health of the climate reflects the health of the planet. Agriculture and cities depend on it. When climate and ice have been in equilibrium, civilization has flourished. When out of sync, well, we are facing the impact now. The story of the climate is the story of the world.

In this light, a 2005 aerial survey of Montana's famed glaciers captured my attention. They all lived in Glacier National Park, but this refuge ironically offered them no protection. Not only had the roster of these alpine glaciers plummeted from 150 to twenty-seven, but scientists believed the retreat of the remaining ice was possibly accelerating. The combined area of all the Park's ice was now 4,028 acres (16.3 square kilometers), down from 5,708 acres (23.1 square kilometers) in 1966 and a fraction of what it had been when I hiked among these mountains and glaciers as a boy.

What had happened to all that ice? What were the implications for the climate? Would the remaining ice fields of Glacier National Park survive?

I was determined to find out. In August 2008, I began tracking the glaciers with Dan Fagre of the U.S. Geological Survey (USGS) in what would become a five-year effort to understand the future of the ice. I discovered that Glacier National Park is a natural laboratory for the prospects of glaciers and alpine wilderness throughout the world. What happens to Montana's ice and the watersheds below it foreshadows the decline of other mountain ecosystems, ranging from the Alps to the Andes. The health of each alpine glacier is a snapshot of the vitality and viability of all frozen summits, a patchwork of ice crowning every continent except Australia. Roughly thirty percent of the land surface of the planet is still ruled by glaciers and their watersheds.

One of the best ways to track the climate, I found, is to monitor glaciers in one's own backyard. No need to go to Alaska or Antarctica. Montana will do quite well. Here, elevated temperatures and decreasing snowfall have cut glaciers to the nub. For this reason, Glacier National Park is one of the best barometers of climate change in North America. Its glaciers acutely feel each notch, each shift in the climate. They mutate. They melt.

This metamorphosis was commonplace during my tenure in the Park. On my travels, I watched the mountains respond to the changing atmosphere the way ice responds to a flame. It began that first August on a glacier named Grinnell. Water poured off the ice field, as if sluices were draining a dam.

While 2008 was Year One of my investigation, twelve months of preparation that year came down to one month of exploration. The schedule was tight. We had only one day on Grinnell Glacier. I soon discovered that Fagre required perfect weather to conduct his glacier surveys—and the best window for dry, sunny days fell between mid-August and mid-September. Each year, he barely squeezed in the work. In fact, Year Two and Year Three of my adventure were often rained out at USGS headquarters in the valley, which corresponded to heavy snows at altitude. Winds in excess of fifty miles per hour often delivered the snowflakes. And there was a consequence. Snow covered the ice, obscuring the glaciers' icy margins and making calculations of the area problematic. Often, we waited around for the snow to clear. Only then could we survey the ice. Glacier is known for its extreme weather; that is how the glaciers developed in the first place. Winter brings the wind at double strength, and the snow drifts higher than a house. However, we had some idyllic moments. Year Four—2011—was the salvation of the project, with many hikes and climbs afforded under sunny skies. Those fair days came with a price, however: The summer sun chiseled away at the glaciers better than a carpenter.

This was my initiation, by ice, to global warming at the local level. I was surprised by the ubiquitous impact of a changing atmosphere. Every inch of the Park was affected. This story is not far afield or distant in time but is happening right now in the United States. Montana, once glacier-rimmed and wild, has been tamed. The white crest of the northern Rockies, what the Native Americans call the "backbone of the world," has turned partially black in summer. The mountains have lost some of their color and some of their grandeur. Still, Fagre and his team are undaunted, with a strong sense of urgency. The following pages tell of their journey, their struggle to take the measure of the glaciers before they disappear, to gauge their history and forecast their future. The team's motivation is tied to a special knowledge: Climate and ice are vital to nature and to civilization. They shape the past and future of the world.

Into the Cirque

Straddling a bright blue crevasse, Dan Fagre stands alone, dwarfed by the ice before him. It stretches for a quarter mile—nearly half a kilometer—in every direction. But the freeze is turning to a rapid thaw. The imposing alpine glacier, launched years ago by a cooling climate, is heating up. Fagre is stripped to a green T-shirt and can feel the warmth—it's another record-breaking summer. Cubes and blocks are crumbling: Small streamlets drain the ice. Fagre, a government scientist, has climbed 3,000 feet (914 meters) to take the pulse of the melting colossus.

Still huge by human dimensions, Grinnell Glacier is one of the last Rocky Mountain giants. Nestled in the mountain cirque, the stadium-like bowl at Fagre's feet, the glacier is enormous—over 150 acres broad and 320 feet (98 meters) thick at the center. However, before man lit a modern match on this continent, it was more than triple that size. Otherwise, the ice field looks about the same as it has since the dawn of industry. Blinding white snow and dull ivory ice cover most of the cirque, from headwall to foot, except for the distant moraines—the exposed rock rubble at the edges. The retreating white glacier has left a brown and barren apron about its periphery. High and low, Grinnell is a study in contrast. Below its foot is silver: Its toe produces a steady stream of glistening ice water that flows over black rocks.

Beneath the ice terminus is an immense lake of meltwater, Upper Grinnell Lake, the glacier's drippings; the lake in turn is vented by another alpine stream that cascades over the lip of the cirque toward the Many Glacier Valley below. From there, the stream joins the St. Mary's River, barreling north through Canada to Hudson Bay, home of the polar bear, where briefly some of the water will reconstitute as ice, before melting again. At both ends of the watershed—alpine and arctic—the ice is slipping away.

I have traveled from my home along the foothills of the New Mexican Rockies to Montana to write a profile on Dan Fagre, the leading glacier expert in the country. For years, I have written about water—topics from sailing to canoeing to diving—but now I return to my first passion: mountaineering. Peaks and glaciers have always meant solitude and freedom to me. More recently, the melting of the ice fields has been troubling: What will the loss of all that alpine water mean? Most of what I've heard about climate change is remote either in time or distance—impacts that are a century removed or that are as far away as the poles. The search for local and immediate manifestations of warming has brought me to Glacier National Park, to learn what I can about our future. I feel like one of those lookouts on the *Titanic*, tracking the path of fractured and melting ice.

The distress calls will come soon enough. Fagre picks up a ball of snow from the crest of the crevasse and lifts it to his mouth. It crumbles in his hand like sand. He blows the last snowflakes into the August wind, and a fraction of them boomerang, stinging his face. The snow in the cirque, he says, given enough time, will turn to ice, and the ice to meltwater, the fresh water joining the sea. Formerly this was a slow geologic process. But the planet is warming at an unprecedented rate. Already, the burning of fossil fuels has elevated the average temperature of the planet by more than 1.5 degrees F (nearly 1 degree C); the heat keeps climbing. Montana snow from the 1960s, converted into ice, is tumbling into the lake. And melting.

Grinnell Glacier is just one of many to suffer. Worldwide, mountain glaciers are on a fast track to oblivion. They are expected to vanish faster than polar ice, both north and south. Grinnell and the other ice-age remnants of Glacier National Park, Montana, may be the first to extinguish at altitude—they are among the most exposed glaciers in North America (and, relatively, the smallest) and thus most susceptible. In 1850, nearly 150 of these glaciers populated the Rocky Mountains of northwest Montana, what would become the Park in 1910. By 1966, there were thirty-seven glaciers or fewer. In 2008, as I tromp around Grinnell, there are twenty-seven. Fagre knows them all.

Dan Fagre (pronounced FAY-gree) is a research ecologist turned glacier scientist. He has been monitoring the Rockies' northern glaciers for nearly twenty years—checking their pulse. He also takes the temperature of the glaciers' snowpack. He measures their dimensions and densities. He gauges their mass. Fagre is a diagnostician. He is the official monitor of the health and lifespan of glaciers in Montana. In 2003, he predicted that, in the face of climate disruption, the largest ice field in the Park (Blackfoot Glacier) would vanish by 2030—nearly thirty years hence. The day he explores Grinnell with me, five years into his forecast, he is reading the glaciers again to see if his timeline is correct. He may have to recalculate.

Within our immediate view are three glaciers—Grinnell, Salamander, and Gem—all of which were connected before a big meltdown split them into a triptych at the dawn of the American industrial age—the advent of factories coughing carbon smoke. Salamander and Gem are poised above the headwall at the back of the cirque like crusted snow on a rooftop. The two overhanging glaciers seem teetering for a crash—each an ice avalanche frozen in time. Stepping back from the gaping crevasse at his feet, Fagre points to these two small glaciers overhead and talks of reading the ice.

"We use sophisticated technology to measure the ice," he says to me, "but we can pretty much tell the health of a glacier by eyeballing it." He points upward, to the left. "Take Gem—that round jewel glistening above the headwall—it's impossible to reach on foot. But from a distance we can tell it's still moving, creeping downhill, because you can see crevasses at its base, just before it slips over the cliff. So we know it's a viable glacier, not just a snowfield."

A living glacier is always on the move, kicking and carving its way downhill.

Fagre traces the outline of Gem and then lowers his eyes to Grinnell, sketching its features with his finger. "Gem is shrinking," he continues, "but the real loser is Grinnell, the mother glacier for this valley. Look at the lateral moraine, that pile of rock and rubble plowed aside by the glacier. The ice is at least 200 yards (183 meters) short of it now—that's how much the ice sheet has contracted in 160 years." Fagre turns his back to me to scan the side moraine. I notice some fir saplings growing in new soil, where ice once roamed, another sign that the glacier began receding a while ago.

While pondering the timing of the glaciers' demise, Fagre often asks why the ice is disappearing and why trees are growing in their paths.

"Trees are migrating to higher altitudes," he says, turning and stooping to pick up a cellophane wrapper from the trail. "It's warmer up here now. Would you ever have imagined a lowland fir invading the domain of a glacier? In mountain ranges all over the world, plants and wildlife are forced higher and higher by global warming. Alpine summits are a cul-de-sac; species are running out of room."

Like a geyser between eruptions, Fagre's passion bubbles over on the hour, like clockwork. I sympathize with his concern. The alpine landscape looks different from my memories of it, hiking here in 1976, when nearly forty glaciers reigned. I remember ice shrouding the mountains like a white powdered wig; now the hairline is receding. I ask him about the widening distance to the moraine, whether that is his best forensic clue.

The ecologist scans the rubble and ravine; it borders the glacier like a skirt. "A broad moraine like this may be old news," he says, "so our best visual gauge of what's happening today is the snow line." This rough line or contour traverses the glacier, showing when there is enough snow to compensate for what has melted. It works like this: In late summer, the visible snow line looks like a meandering hemline with white snow above, which is accumulating, and exposed gray ice below, which is melting. If the snow line is at least two-thirds down the glacier, the ice mass is considered healthy—it is growing or holding its own. Today, the demarcation is clearly up toward the headwall, only one-third down the slope. The rest is wasting away. The official field results: The glacier may be flatlining. In the months ahead I would learn Fagre had a barrage of tests and tools at the ready to profile a glacier. He keeps a diagnostic chart on each, their vital signs carefully listed.

Dan Fagre is a maverick at the U.S. Geological Survey, his employer at the West Glacier Field Station, where he directs the Program for Climate Change in Mountain Ecosystems and a crew of five. He abhors deskwork. Two young men and two women are with him today, presently climbing the last track to join their leader on the ice. Through recent years of a conservative government, he has maintained credibility (and kept above the fray) by undertaking good, unimpeachable science. In return, the Bush Administration has given him a free rein, except in one regard: he is restricted from advocating any specific climate-change policy. By law, he limits his public pronouncements to the causes and effects of global warming—the science. For him, specific remedies and policy initiatives are off-limits. Only political appointees in Washington, D.C., have been allowed to speak. Now, in August 2008, he is perhaps looking forward to the election. But he will not say.

Adjusting to his niche, Fagre has become a jack-of-all-trades in Montana's mountains. As coordinator of glacier-monitoring efforts in Glacier National Park, his activities range from computer modeling to GPS (Global Positioning System) measurements, usually atop crampons or cross-country skis. He is a legend to his crew, most of whom are half his age—he is fifty-six—and they hike and climb at a fast pace, often with Fagre in the lead. He is stocky and fit—built like a mountaineer and rugged looking. His boyish haircut makes him seem younger than his years. But, unusual for a scientist, he doesn't mask his youthful passion. "The loss of a glacier hits me hard," he says. "I like snow and ice. I'd rather be living in the Pleistocene."

The mission of Fagre's alpine program is to study the ecological and geological effects of the global warming trend as it manifests locally in the mountains of northwest Montana. The Park is a crucible, a proving ground for the rest of the alpine world. Global environmental problems often appear here first—the world is watching. Besides recording glacial melting, local monitoring includes reading avalanches, forest fires, stream temperatures (and volumes), and tree line changes, as conditions become more temperate at higher elevations. Not only trees and wildlife are moving uphill. Even coldwater fish are forced upstream as once-cool habitats begin to warm up. Of course, they can't swim much higher—the frigid stretches of streams peter out. They end up in fragmented pockets of cool water. Ecologically, they've been painted into a corner.

Fagre locks eyes with me and clicks a ski pole against his boot, like a spoon rapping a wineglass. He and I stand at the base of the glacier, each with a foot on the ice, the other anchored on rock. He has something to say. "We're not doing ice radar or stream transects today; our objective is strictly to get GPS positions for the glacier—to measure the acreage—by boat and land." That's why he has brought along two rubber rafts. Fagre gently taps the yellow inflatable on his uncomfortable-looking pack. The surfeit of straps on the frame looks like something from the Spanish Inquisition. Just then, one of his crew wanders next to us. Chris Miller carries the second raft and other gear. Fagre continues, "Even a major glacier like this is so small, relative to the resolution of satellite imagery, that satellite remote sensing is not accurate, so we employ aerial photography—and ground-truth anything we get from the air."

Miller says, "That's how I've worn the tread off my boots this summer—ground-truthing for the government. Hauling the gear. Checking every corner of the glacier. I'm a government mule. Lucky it's for a good cause." His boots are caked with snow. A healthy snowpack, after years of winter drought, means it is uncertain if the glacier will recede or grow this year, so Fagre and Miller are even more curious than usual. They plan on mapping the glacier's dimensions from GPS coordinates just to make sure they confirm or disprove the trend. The current hypothesis: extinction for the glacier within twenty-two years or less.

But exactly how much time is really left is uncertain. Grinnell, one of the five largest cirque (or bowl) glaciers in the Park, faces north and, compared to south-facing ice, is slower to melt. Still, it's been losing 2.5 acres annually—on average—over the past forty years. It's in critical condition.

Not all glaciers are equal—massive ones are termed "ice sheets," remnants are named "glacierets"—but all have certain aspects in common. Each glacier, by one common definition, is composed of ice—solid water formed from compacted snow—thick enough (over a hundred feet) for the mass to move forward and downward with gravity under its own weight. The compressed bottom layer is fluid, pliable, and oozes like Silly Putty. It may move inches or feet over the course of a year, but it is always advancing, like a shark. When a glacier is reduced to a size where it stops moving, it is essentially dead. At that point, Fagre cuts it from his viable list.

Grinnell Glacier's age is uncertain: it may date from the last full ice age—the Pleistocene, the time of the woolly mammoth, reaching its height eighteen thousand years ago—or it may be a product of the Little Ice Age, which ended in 1850, or somewhere halfway. From partial ice core samples, the best estimate is seven thousand years. Likely, there has been a cycle of glaciers occupying this amphitheater over the millennia, with dry, warm periods in between. Their size is a balance between growth (from snowfall) and summer melting (called "ablation"), essentially how the glacier advances and retreats. That balance reveals the double threat of climate change: It can affect precipitation as well as temperature. Snowfall has been light over the last few decades. Winter drought and summer warming work at both ends to cut glaciers to the quick.

Drought and melting have been brought on by local warming, reflecting a global trend over the past century. Average global temperatures have increased by 1.44 degrees F (0.8 degrees C) since 1900. (Locally, summer temperatures in the Park have nearly doubled that variation, thanks to its altitude, which nurtures cloud cover that traps heat, and its northern position, which places it in the path of warm air from the Pacific.) This timeline coincides with the overloading of greenhouse gases, from the burning of fossil fuels (oil, gas, coal), into the atmosphere. Named for their tendency to cloak the Earth, these gases allow sunlight through but repress the escape of heat from the earth's surface—like the windowpanes of a hothouse. Greenhouse gases, such as carbon dioxide from car exhaust and electric power plants, are expected to increase steadily, as population and industrialization grow. According to the United Nations, global temperatures may elevate as much as 7.2 degrees F (4 degrees C) by the end of the century. In terms of impact on the climate and weather, this is a huge amount. The warming trend in alpine and polar regions may accelerate even quicker, their temperatures climbing in part because of "positive feedback" loops, such as the tendency of dark open seas (adjacent to floating ice) and exposed mountain walls (next to snowpack) to absorb solar radiation, thus prompting more melting. Then, as more dark water or rock reaches the surface, it brings even more heating and loss of ice. Feedback loops are popping up all over. Consequently, weather patterns will certainly shift. Hurricanes, tornadoes, and typhoons will likely intensify in frequency and force. Rainfall and snowfall will lessen in some places, strengthen in others. In the wake of intense drought, agriculture is expected to suffer in many regions of the world.

While weather is a complex engine and difficult to predict, the melting of glaciers is relatively clear-cut: Heat melts ice. From the Arctic to the Antarctic, and the alpine glaciers in between, as temperatures climb, ice is vanishing at rates far exceeding natural attrition. The North Pole and surrounding waters will be free of summer pack ice for the first time in human history, as early as 2020. The snows of Kilimanjaro will die by 2033. What is most startling is the timescale of the melting: What normally happens over thousands of years is now happening in less than a century. It's as if—after four thousand years—the Great Pyramids crumbled to dust overnight.

Although melting is straightforward, the myriad consequences of glacial extinction are not so simple. Few of today's outcomes were predicted twenty years ago. Surprising interactions include major disruptions to ocean currents from an infusion of fresh water, which in turn will further affect the global climate. More direct links include the rise in sea levels that has already begun. The average global sea level has climbed four to six inches in the past hundred years—so far mostly from alpine glacier melt, though soon to be eclipsed by polar meltwater—and may top three feet or more by the end of the century. More than a hundred million people worldwide live within three feet of mean sea level. As witnessed in the wake of Hurricane Sandy in 2012, New York is at risk, not to mention low-lying Florida, Bangladesh, and the Netherlands. Before this catastrophe reaches full measure with the loss of Antarctic ice, alpine glaciers will melt. Over the century, they will contribute less than three percent of the anticipated sea level rise (because polar melt is accelerating), but their influence will be more immediate on other fronts. Around the civilized world, mountain glaciers and snowpack contribute nearly fifty percent of freshwater drinking and irrigation supplies. Some glaciers are even more essential. On the Indian subcontinent, the Ganges River derives seventy percent of its summer volume from Himalayan ice melt. When the Himalayan glaciers reduce significantly in size, sometime around 2100, India and China can expect widespread drought and famine. For millennia, the Earth has been in balance; in the space of a century, modern civilization and its manipulated climate have torn the equilibrium apart.

On the hike and climb to Grinnell this morning, sea level rise and the exhaust from cars and smokestacks seemed far away. Glacier National Park encompasses one of the greatest natural treasures in American hands. To many, the vistas are as eye-popping as Yosemite. Only wilder. With over 1 million acres of land, the Park is larger than the State of Rhode Island. Historically, it has comprised one of the largest intact wilderness areas in the country. As such, the Park and surrounding land has been termed the "Crown of the Continent." Every creature that resided in these mountains before Europeans arrived still thrives here—predator and prey alike, from beaver to elk to the timber wolf. No major disruption to the wilderness has impinged on its intrinsic value—until now.

Our hike—past mountain lakes, alpine forests, and meadows—was a temporary reprieve. It was easy to forget the global crisis. At a clearing, my attention was drawn to a tiny white Arctic flower, northern eyebright, at the edge of the trail. Distracted, I stumbled into the next spruce, the branches slapping my face and spooking a Steller's jay. Looking up, I was mesmerized by the green-and-blue palette of the near horizon, by the dark cliffs overhead, by the hint of white glaciers around the bend. Midway, the trail hugged a serpentine ridge between an aspen grove and a carpet mat of red-stemmed saxifrage, each mound like a rounded river stone. The gold aspen leaves flashed like shimmering pebbles in that stream. Color cascaded down the ridge. I became lost in the idyllic moment, for a minute listening to two ravens arguing as they soared against the azure sky—a darker blue at that altitude, nearly 7,000 feet (2,133 meters). My whole life, my most awestruck moments, have been offered up by wilderness—days like this. On mountain climbs on three continents, I've marked my path by those encounters, some peaceful and restoring, some violent and raw, but all pristine and humbling. On all those backcountry adventures, man had not been in charge, least of all me. Storms and clouds and weather held dominion. Nature always seemed bigger than us, always in control.

Now, as if to prove the point, a spectacle awaited me around the next bend. A park ranger stood in the middle of the trail with members of a guided tour—all drawing binoculars from their packs—searching the aspens on the slope above us. The ranger looked at Fagre's and my quizzical faces and said, "Grizzly!" Apex

of the pyramid, her kind is always poised between beauty and violence. The great bear had turned her back. Aspen branches crackled
and broke around her, some fifty yards from us. We kept walking,
more briskly now. Grizzlies are the top predator of the northern
mountains, a world of snow and ice. Like the polar bear, they are
unpredictable. This one might have accompanied a family nearby—
mother and cub, a lethal combination for us. I had no desire to see
either one up close.

Sometimes we manifest our worst fears. It will be three years
before I have another close encounter.

Farther ahead, near a stream crossing, we regrouped. The glacier trail was nearly six miles long, rising more than half a vertical
mile from the Many Glacier Valley; we had come halfway. From
the rest stop, we could see the two upper glaciers—Gem and
Salamander—hanging halfway up the flank of Mt. Gould. Our
objective was the larger Grinnell Glacier, still hidden at our angle
behind a high promontory. A waterfall tumbled over the edge.

On a nearby rock I found a series of mushroom-like stromato-
lites, fossils from reefs created by blue-green algae in the Proterozoic Era, over 800 million years ago, when parts of Montana were
covered by an inland sea or ocean delta. Fagre said, "The Rocky
Mountains have only been here for between 150 million and 200
million years, roughly in step with the arrival of the older mountain ranges of today—for example the Andes. Before mountains
started to build, there were no alpine glaciers at mid-latitudes." In
the early Jurassic, it's unlikely dinosaurs often stumbled into ice.
According to one theory, the rising mountain ranges of the Age
of Mammals changed the air circulation patterns of the planet,
making recent ice ages and temperate glaciers possible.

Four hikers happened by. One asked Fagre if he was taking the rubber boat to the glacier to toboggan on the ice, as if we were out for recreation. "A little research on the lake," he answered politely and bent down to pick up another gum wrapper. Litter punctuated the trail.

"Well, we're hiking up to see the glacier before it's gone," she said. This impressed Fagre; he hadn't heard that one before. Maybe park visitors were paying attention. But the other couple was cut from a different grain. The husband of the pair said, "I'm all for global warming—it's too damn cold back in Wisconsin. I prefer Florida." Fagre bit his lip. I followed close behind the leader for the last leg of the trail.

Fagre was born in Minnesota, he told me as we hiked along, where winters were also brutal, though bearing a little less snow. He was a nature kid, even in winter, rigging up wooden cross-country skis to explore and count wildlife. As a teenager, he honed his skills as a naturalist, bird-watching and reading classic natural history, which often had a spiritual bent. Between two extended trips to Japan, where his father taught humanities and philosophy at Tokyo University, Fagre canoed and fished the streams and lakes of upper Minnesota's Boundary Waters Canoe Area, the wilderness made famous by naturalist Sigurd Olson. He also sketched and painted, as an unexpected talent emerged. Wilderness became both a sport and a sanctuary.

Minnesota offered a broad canvas for a boy inclined toward art and nature, but it lacked the logical next vista—mountains. At Prescott College in Arizona, he took to backpacking and soon climbed his first peak at age seventeen—Buckskin Mountain in the Cascades. It was 1970, the first year of Earth Day, and Fagre had found his calling: ecology, the study of species in relation to their environment. This took him to the University of California at

Davis for two graduate degrees. (His doctoral thesis was on coyote behavior.) For the young scientist, alpine ecology would be a marriage of convenience. Science and the high peaks. Vocation and avocation united in the backcountry as one.

At least half the reason Fagre shared so much on the trail was that loud voices deterred the bears. The other option would have been to dangle a few little cat bells from our packs, a common practice, but the joke in Glacier was that you could always tell there had been hikers on the trail by the bells jingling in the bear scat. In any case, Dan kept shouting. During each lull I joined in, with that unnatural volume reserved for yelling at predators and children.

My first climb was Algonquin Peak, the second highest summit in the Adirondack Mountains; I was eleven. My inaugural mountain and the Adirondacks immediately became a temple for me, a holy place in the Druidic sense of the word. I would eventually climb all forty-six high peaks in that vast wilderness—the largest in the continental U.S.—before I turned eighteen. I summited several in winter, I told Fagre, but not Algonquin—a perfect white-capped dome in January. I was proud of my "failure." I knew an older Adirondack guide, Tom Brooks, who always stopped short of a summit and made a short bow, giving each mountain back its mystery. Such a philosophy easily rubbed off on me.

Fagre said, "Climbing peaks, summit or not, is a quest for the aesthetic." He stopped in his tracks and rotated slightly so the profile of his pack and face were in my view. "That's what they hold for me. Mountain literature is full of accounts of the spiritual challenges encountered by climbers, how the expedition fulfills a personal need. Mountains can be mystical, mysterious and beckoning. Who can look at this panorama and not be touched by its beauty?"

In Glacier National Park, the eye-catching wilderness surrounds you in every direction—down, around, and overhead—and swallows you whole. Behind and below us, the Grinnell watershed ran northeast, sculpted by wave after wave of past glaciers that had hollowed out the valley as if a giant hoe had tilled the earth. Flanking the hollow, the rock walls of Mt. Grinnell and Allen Mountain rose from 4,800 feet to 8,600 and 9,300 feet, respectively—a height on par with Half Dome and El Capitan. The ice that plowed this valley must have been over 1,000 feet (305 meters) thick. The unnamed creek, receiving water from Grinnell Glacier, linked together a series of chain lakes—Josephine, Swiftcurrent, Sherburne—that wandered down the valley toward the northbound St. Mary's River. From our perch, the linked lakes below mimicked the giant footprints of some ancient beast.

Those depressions in the landscape, like giant potholes in a prairie, reminded me of the mighty excavating power of the vast Pleistocene ice blocks, now long gone. Carving rock and carrying boulders, by the millions of tons, they pulverized rock and left moraines behind. These huge pools of water stand in their wake. By comparison, the pair of glaciers in our view ahead were mere pockets—ice remnants—from a grander day. Still, I quickened my pace to witness what was left of the Pleistocene and the subsequent Little Ice Age. I was humbled by the sheer expanse of time and space passing by my eyes.

I had balcony seats to an epoch.

Along the valley floor, bordering the lakes, was a spruce forest, but up here we enjoyed alpine meadows, interspersed with aspen and scrub that shielded mountain goats and bighorn sheep, maybe even a wolf, from our view. They'd show themselves soon enough. For now, we kept our eyes trained ahead and our voices cranked up loud—bear-proofing. Around the next bend, we spied Grinnell Falls, a four-hundred-foot cascade, close enough to hear. The glacier lay just ahead, all but its headwall hidden behind the cataract like some lost horizon. Within minutes we would stand on ice. Fagre and his team of four scrambled over the last few steps. The procession slipped into silence, a still reverence, as we entered the cirque.

Stepping back from the crevasse field, Fagre takes in the breadth of the amphitheater, the ice bowl, and the drama in play. Instinctively, he points out features of the rock and ice like an actor reaching for the box seats. At the far upper end of the cirque is a 1,500-foot (457-meter) headwall, a vertical buttress of limestone called the Garden Wall. Its knife-edge summit, called an arête, is the Continental Divide. At the south end of this arête, above Gem Glacier, stands Mt. Gould (9,553 feet; 2,912 meters), a peak that affords a bird's-eye view of Grinnell and the outpost for some of the team's remote photography. Snow falls down the Garden Wall and accumulates on the upper slopes of the glacier, where it is first compacted. Here the ice factory begins.

"Snow gets buried and crunched by the weight above and freezes, then thaws from the warmth of the sun," says Fagre. "Once again it refreezes. It takes two or three years for one winter's snowpack to turn into ice." In 2000, he measured the ice with radar at the base of the headwall. The sheet was thicker than a football field is long. It's slimmer now. From that headwall it takes close to fifty years for the ice to travel, like an industrial conveyor, four hundred yards to the lower lip, where icebergs calve into the lake.

And melt into obscurity. One hundred years ago, beyond the lifespan of this conveyor's load, Upper Grinnell Lake did not exist. Then, the warming took hold. Now it dominates the landscape at the base of the glacier. The lake basin covers sixty acres and is 187 feet deep—all meltwater from the retreating glacier. The ice factory now ends in water, in liquid form—and, for the glacier, it's a losing proposition. Icebergs are quick to melt. Dozens of them float on the lake like outposts for penguins; the scene resembles a miniature McMurdo Sound. The oval lake itself is covered with a thin sheet of "glass ice" and slush; even though it's mild today, the mountain air froze last night. Leads have opened up between the bergs and the surface ice. The water there is colored a cloudy green, its light refracted differently by suspended pulverized rock in the lake—called glacier flour.

Something dark against the distant snow catches Fagre's eyes. At the head of a glacier, a few hundred yards above us and three hundred feet from the rock headwall, a jumble of rock blocks the size of pickup trucks are scattered on the surface of the glacier. These huge boulders fell off the Garden Wall when no one was listening. Already, the glacier has carried them a hundred yards downhill, like a load of groceries. Fagre now has a "marker": The progress of the blocks will tell him how fast the glacier is moving. So far, he calculates it'll take four decades or so for the boulders to spill into the lake—if the glacier keeps moving.

But today, he's more concerned with the glass ice. A thin pane, it coats the surface of the water like film. Fagre and his team of four hauled those two yellow rubber rafts the five-and-a-half miles up here, to paddle across the lake to the terminus of the glacier. From that vantage, where the calving begins, they could reach the glacier's downhill edge to send a GPS signal to the satellites. With the glass ice choking their access, however, the boats are useless today. The team will have to find another way to measure the foot of the glacier.

Even under ideal weather conditions, Fagre has a slim window each year to survey his glaciers. He tries to monitor three or four of them, but must wait until the winter's snowpack has melted at the end of August to see the boundaries of the underlying ice. And he must hurry after that first window opens to register the coordinates of each glacier, to record the loss of ice, before autumn comes. Typically, he is left with the two or three weeks on either side of the first of September to do his job. Summer melting can happen quickly, but the snows soon follow. "We're a little late," he says today. "I didn't expect to find ice in the lake this soon." Dan, if anything, is optimistic to a fault.

Fagre's team gathers around an exposed rock island, about the size of a trampoline, at the southeast edge of the lake. We jump rocks, bouncing from foot to foot, to get to it. His team consists of two young women—Lisa McKeon and Lindsey Bengtson—and two young men—Chris Miller (the oldest) and Dan Reynolds, whom we call Dan Junior. In tandem, they release their packs, drop them to the limestone, and retrieve water and sandwiches. Lisa and Lindsey peel off layers, down to short sleeves, revealing suntans from a summer of high-altitude research. The men (including me) keep their windbreakers on. Except Fagre, of course, who is wearing shorts and a T-shirt. He's an iron man.

There's a slight breeze coming off the lake. Even with the bright sun, the proximity of the glacier is like the neighborhood of an open refrigerator—ice cubes cool the air at our feet. Chris Miller is the last to remove his pack, carefully laying the rubber raft on the rock. He is a burly blond—Nordic in looks and strength. He is the first to speak.

"That was one hell of a portage," he says. "My pack feels like a medieval torture device."

"All that effort and no open water," says Lisa.

"High and dry," Lindsey adds.

Fagre smiles, joining in the chorus. "Ice is a showstopper to-day. I should have thought of it: Dynamite—we should've brought rope, crampons, and dynamite." His face breaks into a wide grin, and the team chuckles. His lament is part frustration, part jest. Fagre wouldn't disturb a glacier with a hiccup.

Lisa, crouching over her daypack at the center of the flat rock, offers her boss some hot tea from her "kick-ass" red thermos. She has bragged on it all summer long.

"Regular or unleaded?" Fagre says.

"Green tea."

"Sugar?"

"Honey."

"My God, no," Fagre sighs. "Give me strong coffee with cream and sugar—the works—every time. Or a glass of something even stronger."

"Sorry, bar's closed, and I left the cappuccino machine in the jeep."

Lisa pours herself a steaming cup of green tea from the legend-ary thermos—remarkably hot for midday after steeping the tea before dawn. Chris kicks a patch of ice from our rock platform into the lake, which breaks through the windowpane ice like an oversize snowball, sending fracture lines across the expanse. Now, with the small snow patch gone, the limestone of the picnic area is completely exposed. Fagre informs his crew that the rock hasn't seen sunlight since the last ice age—as recently as twelve years ago it was under fifty feet of ice. "A dozen summers is all it took," he says. "It's all happening so fast and we don't have much time." He sets aside his sandwich and stands up. He fidgets with his pockets like a man who has lost his keys.

On cue, his team wolfs down half a sandwich each, chugs some water, and buttons up their packs. I stand up stiffly—the long hike has left sore muscles, but I realize Fagre's crew does this thing for a living, so I don't complain. A cloud briefly passes over the sun, and I don some gloves and pull my wool cap down over my ears. The women are still getting a suntan.

The leader steps two paces ahead of the crew. Lisa and Lindsey, Chris Miller and Dan Junior, and I gather round. Fagre scans the ice, which stretches like a frozen field to the headwall bleachers, though triple the size of a stadium. At the base of the back cliff, huge cornices peak above the bench of ice, their tops breaking like white-tipped waves cresting over a placid sea.

Fagre spins around and speaks softly to the team.

"I don't want you climbing near those cornices," he says, "and you can forget those crevasses under Salamander. First of all, they're too dangerous, and, second, we don't need to measure their position. That headwall isn't going anywhere. We have its position pinned down on the computer. It's the other sides of the glacier we have to nail down. But with the lake frozen, we can't use the boats so we'll have to approximate the edges of the glacier." Fagre picks up his two trekking poles. "I want one team— the men—to track the eastern side of the glacier. Climb up the slope and test the snow with a ski pole." He taps the rock with both metal tips. "When you find the edge of the ice underneath, take your GPS measurement—bounce your radio signal off the satellites. That'll be one of three sides of the glacier. We'll get the coordinates of two—your edge and the front. We'll let Salamander go." Fagre points the poles at his crew. "Be conservative." He says slowly. "Don't underestimate the ice."

Fagre doesn't want to give his critics any ammunition.

He peers over his shoulder for a moment at the glacier above us. Then he levels a gaze at the crew. "How does that sound, everyone?"

Chris Miller responds first. "Okay. How many satellite readings do you want for the first edge?"

"Thirty or forty." Chris nods, and the two young men head up the slope.

Lisa says, "I think I'll take some photographs for the website."

"That's fine," Fagre says. "Now Lindsey and I will take the other GPS unit and pace off the lower edge of the glacier, between the crevasse field and the lake. It won't be as accurate as with the boats but we'll get pretty close."

Grinnell Glacier has been measured at least biannually since 2001 (and only sporadically before then)—both directly and with remote photography. The last aerial measurements are from 1998 and 2005, and, because of glass ice, this year will only produce interim data. Breaking with the biannual tradition, the next year will be monitored as well. This data will likely show a ten-year trend—a withering of the glacier since 1998.

By now the men have traversed a hundred yards up the glacier, past three boulders sitting on the ice. Fagre traces their progress. "From where Chris and Dan Junior are testing the ice up that slope," he says, "it looks to me like the eastern edge has receded in a decade—nearly the width of a soccer field. I don't recall those boulders ever being exposed. That's amazing for speed."

Lindsey hoists the bright yellow GPS pack onto her shoulders and buckles her belt. She is Fagre's tech person, trained in GPS research. Fingering the instrument panel, she concentrates on her task, on the minutiae, which is not daunting for a woman who is also planning the intricate details of her wedding this week. She follows Fagre, already two hundred feet ahead, bobbing and weaving through the crevasse field again. The crevasses are bunched together. While the rock slope beneath the upper half of the glacier must be relatively smooth and gradual, here—at the lower end—the underlying bedrock is likely a tumble of rock. The foundation may bear a hidden precipice over which the ice cracks and tears. It cascades like a frozen waterfall. The violence causes splits and fissures in the ice that widen into deep crevasses, which penetrate far down into the glacier. At the edge of the lake, huge chunks of ice already vivisected by these crevasses breach from the glacier and "calve off," floating free as icebergs in the lake.

Fagre and Lindsey stay clear of the calving zone, skipping and hopping as if playing hopscotch over smaller crevasses near the uphill fracture area. I join them, gingerly stepping over each crevice, from snow edge to ice, from danger to safety. I would prefer to wear crampons on my feet and have an ice axe in my hands. The crew climbs without a net. The ice ramp is dirty, littered with rocks and rubble and debris carried from the headwall as the conveyor grinds away. We climb over, around, and through an especially dark series of crevasses that border a gaping abyss. The cavern may be 150 feet deep. Be careful here, I think. This is not the Khumbu Icefall on Everest, but deadly nonetheless.

In slow motion, Lindsey hurdles the crevasses—crossing them perpendicularly—with startling agility. She is a brunette, with long straight hair that flops each time she jumps. I couldn't match her coordination—she was the strongest hiker on our climb earlier in the day. And she is patient with the older men. She learned her etiquette while serving as a wilderness guide in Glacier. Lindsey negotiates two more crevasses and turns around to check in with the leader. A short, yellow pole with a yellow disk on top, like a saffron mushroom, protrudes from her pack—the antenna for signaling the satellites.

Fagre says, "A few more steps, closer to the lake."

Lindsey looks dubiously over her shoulder. She is dangerously close to the calving zone but doesn't complain; she has trained for this for several years. Lindsey retreats another six feet from us, carefully over broken ice, and lifts her eyes.

"It's okay if we measure the glacier too big," Fagre calls out to her, in response, "just not smaller than it is. We've got to be cautious in our predictions." He surveys the scene to the left and right (and behind Lindsey), and acknowledges all the obstacles around. He's a perfectionist but protective of his crew. He holds up his right hand, with index finger and thumb barely spread to indicate a smidgen. She withdraws half a step more.

"Okay, that'll do. Make your reading."

Lindsey grabs the handheld yellow GPS receiver and pushes several buttons. In less than a second, the pulse bounces off three satellites orbiting the Earth. Within the same second the signals reappear on her instrument panel. Technology has caught up with the ice age; here time is compressed on every level. She has a precise reading for the foot of the glacier in the blink of an eye.

The Global Positioning System, developed by the U.S. Department of Defense, is a worldwide radio-navigation scheme that registers an object's exact position, within one or two feet. Lindsey's receiver measures position and distance by analyzing the differential travel time of radio signals to and from three "close" satellites. Through "trilateration," using the geometry of triangles, the exact position of the antenna pack can be located. This tracking relies on each radio signal always traveling at the speed of light, and therefore the travel times to and from each satellite change as the antenna moves about. For example, when the edge of a glacier retreats, it may move closer to a satellite, making the travel time of the radio signal shorter. A clock records these times. Thus, the distance of the antenna from each satellite can be calculated, pinpointing the edge of the glacier. The processor on the instrument panel of the GPS unit assigns a latitude and longitude to the position. Back at headquarters Fagre and his team plot these coordinates to map the glaciers.

Fagre stares into the crevasse between Lindsey and him, lost in thought for a moment, then lifts his eyes skyward in the direction of the satellites that will read the health of his patient. "In seventeen years," he says, "we haven't had one glacier advance—or increase significantly in size—in the Park. They're all wasting away—maybe faster now than ten years ago." He was losing two-and-a-half acres of ice each year; now it's over three. The glacier may be leaving all forecasts, all predictions, in the dust.

We talk about the causes of climate disruption for a minute, the global picture, a conversation more suited to New Jersey than Montana. Perhaps. We have too may cars, too many coal-fired power plants, he says, which are being built weekly in the developing world and still operated with little carbon recapture in the United States. At this point, even remedies like carbon-neutral power plants can't save our local glaciers, he says. The sky holds too much carbon now. Fagre is less a doctor than a hospice worker; the loss of Montana's glaciers may be beyond his control.

"We're approaching a tipping point," he says, "beyond which there is no turning back. At that point, cutting our own emissions won't help—it can't reverse the damage. I don't know how much time we have. Some say ten years. But at some stage, on a warming planet, so much water vapor will evaporate from the oceans, adding to the greenhouse layer, that heating will accelerate even more and there'll be no stopping the temperature rise. We'd have runaway climate change. If that happens, even the Himalayas and the poles could melt."

The second team—Miller and Dan Junior—have climbed farther by now, to an ice gulley at the eastern edge of the glacier. I traverse the ice to take a look. Here, they prod the snow with ski poles to locate the underlying ice. The probes strike the frozen layer, resounding on each tap like a pickaxe hitting rock and echoing off the cirque with a *ping*. The edge is closer than they thought. Fagre ascends the glacier diagonally to join us, and together we inspect the lateral edge, where the ice has pulled back from the moraine. The architecture of the ice is in full view. From the side, the ice is hollow underneath, like the cross-section of an igloo. Fagre is floored. He shakes his head in disbelief.

Lindsey finishes her readings and joins the men, next to a square boulder carried from the headwall by the ice conveyor. She marvels at the ice cave. "This part of the sheet is riddled with tunnels," she says. "That's where much of the melting must be happening. A river is running underneath, carving out the glacier."

Over the next year or so, back at the office, the GPS measurements and subsequent mapping will show that the terminal edge has retracted 12 to 820 feet (up to 250 meters) since 1998, in ten years' time. (The largest wastage is near the lake.) The glacier has lost an estimated 30 acres (0.12 square kilometers) in surface area, or seventeen percent, but its vertical mass could have plummeted even more. Today's disturbing evidence of a hollow core confirms this likelihood. Next summer, Fagre plans on employing radar again to gauge the current depth of the ice. He will likely see that the front profile near the lake is narrowing, too, but by how much? He'd have to be patient as Job. In the meantime, the glacier would likely diminish—it was structurally unsound, so more calving was likely. More icebergs would certainly break loose and drown in the lake.

We glissade down the gulley to join Lisa at our luncheon rock. The snow is slushy after a day in the sunlight, and the ice pops into view, surfacing in gleaming patches like a row of miniature mountain peaks. I think of the hollow dome underneath my feet, a canopy like the ice atop a winter lake. It reminds me of those ice-capped basins in the Adirondacks of my teenage years. Only now I am not interested in going ice fishing. Or diving. We hug the lateral moraine and I feel more secure. If I crash through, I can always jump toward the moraine.

Lisa has made some beautiful photographs, covering the glacier from head to toe. She is Fagre's memory, taking the thumbprint of the giant. She has to act fast. Like a metronome, the clock of the leviathan ticks away. I scan a few through her Nikon viewing monitor. She has captured both the headwall and the height of the icebergs by lying prone on the shelf at the toe of the glacier. Mountains and glaciers are often foreshortened in photographs, but she has given them their majesty.

To reconnect with the trail for our descent to the parking lot, we have to skirt the lake, its full breadth, crossing over the exit stream. This course, with attendant boulder hopping, gives us a better vantage to see the front of the glacier, where it has cracked and crevassed and calved into the lake. A dozen bergs huddle at the convergence, each tip rising over twenty feet high, their greater dimension—nearly 140 feet—underneath the water surface. They are bright white, like polished ivory, and reflect the afternoon sunlight sharply, making my eyes squint.

We stop on a small promontory, gathering around our leader. Fagre says, "Al Gore hiked up here with me in 1997. He wanted to see one of our glaciers firsthand. Up to that point, the lake had just been meltwater. Very little ice floated on its surface; it was clear. Then, two days before we arrived, there was a huge calving event: The leading edge of the glacier broke, splitting off a large chunk that shattered into dozens of small bergs. Ever since Al Gore's trip, the lake has been choked with icebergs. He saw some of the first days of the disintegration of Grinnell."

Curiosity about the disintegration of ice worldwide led Dan Fagre to Mt. Everest base camp the previous fall—to see how the world's largest mountain chain was faring in the growing heat. He witnessed the retreat of the Khumbu Glacier, Fagre tells us as we huddle on the hill, and mourned the damage to Earth's greatest peak. He also made the pilgrimage to feed his mountaineering hunger. Trekking the Himalayas was the fulfillment of a lifelong dream. It is Fagre's core belief that humanity can save Everest's glaciers and their neighbors.

"But my most cherished mountain, my favorite climb, is not the highest summit," he says. "It is Glacier Peak in the Cascades. It's a beautiful mountain—an ancient volcano that now comes sharply to a summit. I climbed it in 1978, one of my first American peaks. It was a perfect bluebird day. My partner and I were alone on the mountain. With all the towering ice, we felt like characters out of James Hilton's *Lost Horizon*."

Glacier Peak (10,541 feet; 3,213 meters), which sits in the North Cascades of Washington State, is adorned with Scimitar Glacier (Fagre's route), which winds like a curved blade down the mountain, and four other large ice fields. Fagre says they are still there. I climbed the peak, too, via Kennedy Glacier in 1986. I recall the summit rocks, a labyrinth at the crest of the glacier. "I slipped on one of those rocks and cut my leg with my own ice axe," I say. I show him the scar.

We trade knowing looks.

Thus began our tradition of exchanging mountain stories, like two shipmates talking about old storms at sea. "Glacier Peak was my third or fourth outing on a glacier-clad mountain," he says. "But it was better than a first date." That early on, I tell him, none of us had yet figured out all the right moves.

Dan Fagre, it turns out, is a romantic—at least as far as mountains go. And memories of mountains usually run deep.

By now, the glass ice has opened up in places, giving way to open water. It is 3 P.M., however, too late in the day for boating. We have to head down before it gets dark. Not only is the window narrow for research each season, but, with a long approach hike, the day is short as well. "We'll bring the boats up again next Labor Day," Fagre says, "and paddle up between those bergs to the front of the glacier. Then at last we'll have a precise location of where the glacier lies in its bowl."

Lindsey and Chris pick up their backpacks with the yellow satellite transponders and, hoisting them to their shoulders in mid-stride, tramp down the path. Lisa and I wait for Fagre, who adjusts the straps of his medieval torture device, not that it will do any good. Dan Junior wanders our way, kicking the dirt. He has something on his mind.

"There must be some way to save these glaciers," he says, "some way to reverse the clock." Dan Junior is the newest and youngest recruit to the team, on loan from the National Park Service. He has not yet logged in enough time to become realistic about the Park's chances.

Fagre stares at his young charge for a second, then lifts his eyes to the icebergs beyond. He speaks hesitantly, like a man choosing his words carefully. "You know, the government limits me on what I can say. I can't say everyone in the country should demand cars with fifty miles to the gallon or demand compact fluorescent lightbulbs. I can't say that. The Bush Administration won't let me. I can't tell people what they should do. I can only say options exist to cut down carbon emissions, if that's what you want to do. Is that what you want to do?"

"Of course."

"Well, then, you need to stop burning fossil fuels."

"What's the best way to do it?"

Fagre shakes his head in obvious frustration. His glaciers are in critical care. He knows all the antidotes, but can't get government approval for the remedy. He slips into an uncomfortable silence. Perhaps to distract himself, he picks up another gum wrapper—refuse of the trail—and stuffs it into his shorts.

Fagre's brief discourse with Dan Junior is two months before the Obama–McCain election. From where his team stands, it is years away. However, Fagre doesn't believe a new president will necessarily be a quick fix.

"Things will be different after the election," Dan Junior offers.

"Maybe," Fagre says. "However, curbing climate change always comes down to economics. What politicians don't realize is that the environment is our greatest resource."

Chris adds, "You can't have water without the snowpack. You can't have medicines without the wilderness. Or honey in your tea."

"Now you're talking," Lisa says.

Fagre smiles, then shuts his lips. He is not yet free of his shackles.

Before departing, the four of us scan the Grinnell amphitheater one more time, drinking in the beauty. And the tragedy. I scan the six hundred vertical feet from the tabletop of the glacier to Salamander's ice apron above, a sheer precipice with a brilliant trickle of water draining the edge. I realize Grinnell once stretched up to embrace that high sparkling ice—yet she has only one-third that reach today. It's as if she has been withdrawing into herself—retracting and retreating—a magician that vanishes into her own white cloak.

I kick a stone into the lake and the wandering ripples, staring at my feet, race to mid-channel like a gyre and disappear under the glass ice. The wind now dapples the water—just at the edge of freezing. To some a wasteland, to me the alpine zone above tree line seems alive, resolutely alive—each animal, each shrub, each ice crystal holding a niche where others would fail. That is the power of the place: living on the edge, at the limits of existence. Nature is stripped down to its essence here, and yet it prevails—apart from us. Nature doesn't need us, yet civilization is totally dependent on the environment. The flow of services is one way. Yes, glaciers offer direct benefits to humanity—drinking water, irrigation sources, recreation—but it is all the indirect and subtle traits that touch me today. Like the unexpected bright blue of a crevasse. I am humbled by this glacier. I am in awe of her. Above all, a mountain offers us humility. A little less hubris may save the climate,

At my feet I discover an alpine glacier poppy, a single, black-centered, orange flower barely rising out of the sharp, red argillite scree. Its existence seems improbable in such a harsh environment. This pygmy poppy is endemic, meaning it grows in only one place on Earth—here, specially adapted to the alpine tundra of Glacier National Park. The lone poppy stands up to the wind like a mast in a storm. With its local climate shifting, its singular habitat shrinking, I wonder how long this rare species can survive. Its essential cool temperatures are wanting. The flower is rare and alone, yet it touches me. We are all in this together, I consider, and bow my head to authenticity.

I rejoin the group. Lisa takes one last picture: three men with red parkas on the edge of an ice floe, as if we were polar bound.

"I prefer the mountains," Fagre says. "Something about the heights stretches your mind. Before I was forty, I climbed a bunch of peaks on this continent—Mt. Baker in the Cascades, Popo-catepetl, that huge volcano in Mexico, six Bugaboos in British Columbia—and I always thought the glaciers of North America were secure. No indication they'd vanish, at least not on this scale—maybe in a thousand years but never in my lifetime." Fagre walks down the trail, stooping here and there to pick up more trash—cellophane bags and a fiberglass antenna discarded by a previous team.

"Now everything's insecure," he continues. "Just as disappearing birds signaled a silent spring fifty years ago, vanishing glaciers are a bellwether of climate change. They show the health of the planet. Glaciers are measuring the crisis, a huge crisis—something that's going to fundamentally alter civilization."

We can't catch up to Lindsey and Chris—too far ahead and too young—but perhaps they will spook the grizzlies. Just the same, we speak loudly on the descent, forcing the conversation after a long day. Fagre has recovered from the rant about public policy, his attention returning to his twenty-seven glaciers, the jewels in his crown. Grinnell is his touchstone, one of the few he visits almost every season. He tells Lisa, his senior assistant, that they'll have to work up today's coordinates as soon as they get back to headquarters. He is anxious to plot the glacier's margins, to make a prognosis. In his mind, he is suspicious the melt rate will prove worse than two years ago. The hollow edges suggest it. But he is always meticulous with the data. Science should remain separate from policy, he says, his heart with his profession, his mind with the future.

"We may have to revise the 2030 forecast," Fagre now shouts, as much to the grizzlies as to us. He stoops to pick up a discarded penny. "I'm not sure these glaciers have twenty-two years left." If the glaciers do vanish sooner than 2030, Glacier National Park may have to give up its name. Fagre pauses to look back up the trail at the triptych of glaciers, then stands and spins around to face us. He tosses the penny high—it hangs in the air for a moment then lands in his palm, but he conceals the outcome.

"Yes, sir," he says with the coin in his fist. "The whole world is turning upside down."

Through the Looking Glass

Near the western entrance to Glacier National Park, a row of five wooden cabins appears to be hibernating, the line of green tin roofs burdened with icicles and snow. A few snowflakes dance in the early morning light, which slices through the trees at the sharp angle of a boreal forest at winter daybreak. Footprints to each door and smoke from the woodstove chimneys betray the apparent repose. Inside, government employees scurry and process data from the previous summer's season in the Park. The middle cabin of the five is the headquarters of the Center for Climate Change in Mountain Ecosystems (CCME), base camp for Dan Fagre and his crew, especially in winter when deskwork doubles. The snug building holds computers and survey gear, along with backpacks, snowshoes, and cross-country skis. In winter and summer, glacier expeditions are launched from here, from what the CCME team calls "the lab."

We have jumped forward a year to December 2009, and weather across the United States has been brutal. It will be a long winter. The worst snowstorms in decades will hit the East Coast come February. The Midwest and Southwest will be pummeled, too. But El Niño, demonstrating its split personality, has delivered a winter drought to the Northwest. The jet stream has severed in two, forking north and south, nearly bypassing Montana. Temperatures are moderate. For the peaks of Glacier National Park, this has hinted again at a new pattern for winter weather: For the most part, the coldest months have been bone-dry and mild.

Dan Fagre peers out the window next to his computer, searching the light snow for signs of what's to come. In his hand is a black phone; he has me on the line. Fagre mouths a couple of words to Lindsey, his "tech person," and tells her to wait a minute. He pours his third cup of coffee. Then he turns on the speakerphone and sets the headset in the cradle. He roams the room. A man that cannot stand still. Or stop worrying. He acts like someone who is running out of time.

"This dry season is bad for the glaciers," he says to me. "Bad for the mountains, bad for the rivers. We need a good dump or two before spring to build up the snowpack. Otherwise, with a warm summer on top of it, we're going to lose more glaciers. That's bad for everybody downstream."

The modest winter weather has been a growing trend, but not officially a full climate disruption as yet. Only a multidecadal pattern, a long-term temperature or precipitation shift, signals a change in climate. A local climate comprises the average atmospheric phenomena over several decades, at least thirty years. Whereas, weather is what happens next week—or last. The distinction is lost by some skeptics of global warming, who look at recent heavy winters in the eastern United States as proof that global warming is a hoax. They see the cold and snow as a litmus test for what they claim is really a cooling period. However, one or two winters doesn't make a climate. It would take several decades of super-cold winters and cool summers to reverse mercury's upward progress worldwide. In any case, the following summer will likely be a record, too—a hot one all over the United States. In Glacier, the number of extreme hot days will triple. That's what climate change models predict: greater extremes of weather at both ends.

Another shift has been the tendency of annual precipitation in Montana to fall more often as rain. Formerly, the state's water supply was dominated by snowfall; now rain is a more likely water-bearer. This has wide consequences for mountain ecosystems. With less snowpack, water reserves disappear by midsummer, making for a dry autumn. Also, drier conditions in spring and fall extend the likely fire season. Rain dropping on snow brings on avalanches, greatly affecting the mountain landscape. In all, there are dozens of impacts of the vanishing snows. Two or three heavy winters, of course, could moderate the trend.

Precipitation is not only an issue in winter. The success of each summer survey is at the mercy of the rain. During my tenure at Glacier National Park, the first few summers have varied dramatically in weather and accessibility to the ice. For example, Year Two (2009) was warm and dry—with a successful trip to Grinnell Glacier—while Year Three (2010) was cold and wet. We slogged our way to Grinnell with a CNN News team and were nearly blown off the glacier's high moraine by rain and high winds. It would not be until my fourth year, 2011, that all the conditions were right for a trifecta—visiting three glaciers in a single season.

But right now, on the phone in December, winter is on my mind. Dan Fagre tells me his desk is littered with papers, his graphing calculator loaded with numbers, the tumblers ready to fall. These are the makings of an inventory of the glaciers, and an assessment of the ice, for the Park's centennial in May. He is relying on aerial photographs to compute the acreage since they are available for all the Park's glaciers. The latest shots are scattered on furniture around the room. "We're comparing aerial images from 1966 and 2005 to measure the glaciers," he says, while sipping his coffee. "So far, every one of them has retreated, some by less than ten percent, some better than fifty percent. I'll have to get permission from USGS headquarters in Reston, Virginia, to release the actual figures."

Fagre still has constraints in his government job, but the chains have loosened somewhat. With a new administration—the Obama White House—his annual budget has been more secure. He's been freer to plan the field schedule and to speak. He's been more at ease with reporters. Climate change, if not on the front burner nationally, is not at the back of the stove either. And its center of gravity is in Glacier National Park.

The snow has stopped. The sun has climbed slightly along its gentle arc across the southern sky, bringing the face of the cabin briefly into sunlight. Fagre tells me the roof below the chimney has fallen into shadow. Like its brethren on cabin row, the CCME building is part of the GNP Headquarters Historic District that is listed in the National Register of Historic Places. The cabin was designed in 1917 and stood as a sentinel at the early gateway to the Park.

The Park was established only seven years earlier, on May 11, 1910, and will celebrate its centennial with much fanfare. The legislation set aside over one million acres (in excess of four thousand square kilometers) of forest and meadows, parts of two mountain ranges laced with ice, and over 130 glacier-carved lakes. The mission was officially spelled out six years later: ". . . to conserve the scenery and the wildlife therein. . . . by such means as will leave them unimpaired for the enjoyment of future generations."

But the history of the protection of these lands and resources goes back further, to 1897, when the Great Northern Railway, after laying tracks across the Continental Divide at Marias Pass (5,213 feet; 1,589 meters), successfully lobbied Congress to designate the area a forest preserve. This action opened up the region to tourism and development. After the creation of the Park thirteen years later, at the behest of George Bird Grinnell, an early explorer and promoter for whom Grinnell Glacier was named, and other enthusiasts, the construction began. Grand log hotels and lodges were built by the railroad barons to capitalize on the growing summer traffic.

The tourists arrived to see the splendor—the elk, mountain goats, bighorn sheep, and over three hundred grizzly bears at the apex of the ecosystem. Yet the primary attraction was nearly one hundred glaciers radiating like fingers and thumbs kneading away from the bases of the mountain peaks. Even the lakes and valleys along which visitors lodged in luxury had been chiseled and sculpted by ice over thousands of years. The tourists came for the living glaciers but also for the rock sculpture from the Pleistocene.

A chance event some sixty to seventy million years before the Great Ice Age—that is, before the Pleistocene—primed the mountains for the shape-shifting to come. This is when a great rock wedge traveled east and landed at Glacier's doorstep. But the full story of Glacier's rocks goes back another billion years or so to the

Proterozoic (Precambrian) Age, when Montana had no high mountain ranges at all. In the northwest corner, the region was as flat as a prairie. At that time, it is thought, a great basin—called the Belt Sea and looking perhaps something like the Caspian Sea—dominated the otherwise barren landscape about fifty miles west of the present-day Park. Rivers fed the immense sea, carrying mud and other silt rich in calcium carbonate and quartz. The sediments fell to the bottom of the basin in variegated layers. Evidence persists to this day. In the shallows, waves rippled the bottom mud; raindrops and hail dented alluvial sand on shore. Those ripples and raindrops are preserved in the rock that sprung from the sediments. I find such dimpled rocks on my hikes around the Park.

The Belt watershed must have looked like Tornado Alley. Without land plants to curb erosion during a rainstorm, enormous sheets of water barreled across the landscape, scooping up stones and sand, as well as mud. Debris littered the seashore. Today, Belt sediments become finer as we move east, with stones dropping out first and sand last. This suggests that major rivers originated west of the basin because heavier, coarser deposits tend to settle out first when rivers lose their power. This gradient provides strong evidence for the basin being an inland sea. If it had been a coastal bay instead, then no rivers would have run from the west. Just the Pacific would sit there. But geologists believe another continent adhered to the west coast in the Proterozoic (more than 542 million years ago)—beyond the inland sea—most likely one comprising a good portion of Siberia. Siberian granite matches the sand grains embedded in Belt rocks.

Wedged between two continents, the inland sea was essentially a rift like the Great Rift Valley in Africa. But like an imperfect weld, this joining was the first to split apart as tectonic plates diverged. For the time being, however, through the middle Proterozoic, the Siberian rivers likely flowed east, spilling their silt into the alluvial fans and depths of the great basin. Over hundreds of millions of years these deposits sank under their own weight, formed solid bands, and under pressure and heat, transformed into rock.

Eventually the basin filled. Sedimentation ceased. The sea dried up. The newly formed sedimentary rocks—limestone, argillite, dolomite, and others in thick layers—sat undisturbed for nearly one billion years in the cradle of the old seabed.

By then, by the middle Jurassic (180 million years ago), dozens of islands wandered the Pacific; some archipelagos the size of Japan crashed into North America, one after another like a series of freight trains colliding. The effect of the smashup was the crumpling and vaulting of the continental crust into long chains of mountains like the early Rockies. Even after the islands ceased colliding eighty million years ago, the compression continued. The Rockies rose further, building for another thirty-five million years.

In the blink of a geologic eye, this same compression fractured a section of the basin's rocks, thrusting a chisel-like plate three miles thick and 160 miles long eastward across northwestern Montana. This was the famous wedge. It heaved up and over the younger Cretaceous rock, burying it and exposing the Proterozoic layers—over 1.4 billion years older—on top. It was geological topsy-turvy, like turning over a rock garden. Part of the exposed, ancient block would become Glacier National Park.

Geologists would name the great rock wedge the "Lewis Overthrust," after Meriwether Lewis, who explored territory just to the south. The slab took millions of years to move uphill to its

present address, at a speed of approximately one inch per year—
about the rate of growth of a fingernail. The exposed sheet presents a dozen layers from the old basin sediments. The deposits
are arranged in horizontal cross-section like the dirt and clay and
cobble along a stream bank, just on a different scale. The strata of
the Belt Formation span over 15,000 feet (4,572 meters) thick with
alternating color bands ranging from plum to pistachio to orange.

One of the most visible examples of this inversion can be seen
near Marias Pass on the southern approach to the Park. In the
distance, halfway up Little Dog Mountain is a clear horizontal
break—a fault line. Below this gray band is seventy-million-year-
old Cretaceous rock; above the band lies limestone over 1.6 billion
years old. The younger rock holds up the ancient.

Youth has its burdens.

The limestone band at Little Dog Mountain is called the Altyn
Formation and is one of the earliest foundations of the great
wedge, the overthrust. Composed of tan dolomite and limestone,
it often sits upon the younger Cretaceous underlayer like an up-
side-down cake. Just above is the Appekunny Formation, ranging
in color from bright green jade to dark olive. The green layer bears
fossils of an early metazoan, *Horodyskia moniliformis*, appearing as
if it were a string of beads and considered one of the earliest animal
fossils on Earth. Its discovery pushed back the official date for the
origination of animal life on the planet by a billion years.

The great wedge provided the raw materials for the extent and
contours of northwest Montana's Rocky Mountains, but the actual feat of sculpting them came millions of years later. Ice was
the carver extraordinaire. And it was fast in its handiwork. It took
much longer to build the mountains than to whittle them down
to size.

A glacier's ice is a rare commodity. Like the diamond it resembles from a distance, its origin depends on just the right circumstances. While each glacier owes its life to its first snowflake, the ice grows from billions of them. These crystals must combine and compress, melt and refreeze, to mutate into ice. A warm sun and cold nights are ideal for this transformation. This was the likely state of affairs when glaciers first appeared in the Rocky Mountains.

Remarkably, the Rocky peaks and their alpine brethren worldwide may have prompted the conditions for their undoing. The rise of mountain chains was possibly one of the stimuli for changing atmospheric conditions that ushered forth recent ice ages. According to this theory, the growing height of mountain ranges altered the circulation of cold air and moisture. The new currents may have brought more snow and inhibited melting. Glaciers would have grown and advanced. The high mountain profile may have set such a stage for some of the ice ages—possibly dozens of them—in the past 100 to 200 million years, but there were likely a combination of forces behind each climate shift. Scientists have suggested that continental drift may have partly spawned early ice age cycles, particularly the Permo-Carboniferous Ice Age (300 million years ago), again by disrupting global weather patterns as continents shuttled back and forth. (For much of the time, the supercontinent Gondwanaland was centered near the South Pole and burdened with ice.) Another theory, which has special resonance for the most recent glaciations, is the idea that variations in Earth's elliptical orbit influence the solar energy received and thus the climate expressed. The Earth not only is eccentric in its rotation about the sun, it also wobbles—what is known as the procession of the equinoxes—in a 23,000-year cycle. Halfway through the cycle the tilt of the Earth will be opposite what it is today, changing daylight length for any given spot on the planet. The tilt and eccentricity might trigger a climate response. Best correlated: The eccentricity of the Earth and ice age occurrence both follow a longer 100,000-year cycle. This is a better fit than the 23,000-year wobble. But while the astronomical concept is gaining adherents, none of the hypotheses have yet attained the rank of an accepted scientific theory. Why astronomy and climatology follow the same metronome is a mystery.

But there's beauty to it. The celestial correlation is likely more than a coincidence.

By three million years ago, glaciers arrived periodically in waves. These pulses comprise the Great Ice Age (the Pleistocene), during which ice advanced across North America four times, each foray carving and sculpting the Rocky Mountains. The most recent of the four, the Wisconsin Glaciation, was punctuated by multiple mini-advances (colder periods) and recessions (less cold), reaching its pinnacle about eighteen thousand years ago, when it capped a third of the continent with ice. This blanket lowered sea levels by up to 300 feet (91 meters). The Wisconsin Glacial administered the coup de grace to the Park's mountains, dismantling much of the construction of the great wedge.

During the Great Ice Age, upon each advance, snow often fell year-round, feeding the glaciers with the raw ingredient of ice. At the spine of the Rockies—today's Continental Divide—mile-thick glaciers covered all but the tips of the tallest peaks. The rock summits breaking above the ice would have appeared as an island archipelago in a sea of white to the stone-age hunters of the adjoining plains. To the south of the present-day Park, the land was ice-free.

Giant bison, camels, and mammoths grazed on the tundra and shrubs. Wolves, cheetahs, and saber-toothed cats hunted the plains. The oldest human bones from this area date to 10,600 years ago (plus or minus 300 years), and tools from nearby sites date several thousand years older. Early man lived at the edge of the glaciers, not unlike Fagre and his team today.

Hidden below the placid surface of the ice sheet, frozen water chiseled away at the mountain flanks. The glaciers were busy quarrying rock. The excavation was on a grand scale. Thousands of square miles were pummeled. Millions of tons of limestone and other Belt rocks were severed from the mountain walls and carried downslope with the marching ice. A mile-thick glacier can dislodge, pick up, and transport rocks as big as a house—with ease. As miners of stone, glaciers have only one equal—the planet's tectonic plates. While ice cannot displace a continent, as the plates can, it can whittle a mountain down to a hill.

Glacial ice eroded the landscape in two ways: plucking and abrading. First, the glaciers loosened and plucked up fragments and stones, carrying pebbles and boulders as big as buildings. The plucking process often began with pressure melting, in which a thin film of water on the bottom of the glacier seeped into fissures in the bedrock. As the water froze and thawed and refroze—expanding and contracting and expanding again—the rock fractured. The glacier then plucked up the chips and chunks and hauled them away. Like a pick and shovel, the glacier could also pluck loose random boulders and huge broken shards, clearing a path. Sometimes the glacial ice simply froze to a stone, adhering like glue and whisking it away. Abrasion was the second major technique for slicing up rock. Stones embedded in the underside of a glacier acted like oversize garden rakes, scratching grooves in the bedrock. These grooves, called "striations," left purchase for further plucking.

On the west side of the Park, a trunk glacier, so called because it was a main stem with branches radiating from it, descended into Montana's North Fork Valley from the mountains of British Columbia. The trunk flowed like a river, ferrying Canadian debris for several hundred miles. Abraded boulders looted from those Canadian peaks now sit atop Huckleberry Mountain, over 3,000 feet (914 meters) above the North Fork River. The Canadian boulders rest on striated bedrock from the giant wedge that dominates the Park. Once, I ran my hand across these striations. They felt like grooves on a banister, the long sweep from attic to front door, from past to present. Millennia in the palm of my hand. To me, geology makes time visible.

About twelve thousand years ago, with a change in climate—a warming trend that drew the most recent glaciation to a close (perhaps astronomical in its cause again)—the glaciers began to melt. As the ice trickled away, the artistry of the glaciers was slowly revealed: horn-shaped peaks, bowls and hanging valleys, knife-edge walls, and moraine-encircled lakes. The sculpture was on a colossal scale, and the unveiling took centuries. But even as the Great Ice Age ended, Montana's Rockies had not seen the end of ice. There were occasional modest advances, followed by interglacial periods and then by more minor advances. (The current glaciers formed perhaps seven thousand years ago—sometime after the Great Ice Age, during an interim cold snap.) Then, well into modern times, came the Little Ice Age (1300–1850), in which the glaciers advanced during 550 years of colder temperatures in the Northern Hemisphere. When the cold snap ended, the ice of

150 glaciers speckled the land of the future Park, only sixty years before its founding.

The Little Ice Age was a climate swing rather than a full glaciation, and the causal agents were likely different. Compared with a bona fide ice age on a 100,000-year timescale, the short-term expansion of glaciers was prompted by only a small change in temperature, no more than 1 or 2 degrees Celsius (less than 3.5 degrees F). In climate science, cause and effect are often difficult to discern; however, scientists like Dan Fagre believe that the oceans played a central role in initiating and maintaining the Little Ice Age. Because oceans store heat from solar radiation and disperse this energy through currents and gyres, they drive much of the world's climate. If the oceans' average temperature is elevated, then so too the atmosphere. For some unknown reason—perhaps low sunspot activity—solar radiation may have decreased during the Little Ice Age. Then, it rebounded around 1850, as the period came to a close. Perhaps the waning of the Little Ice Age was also helped along by the burning of coal and other fuels in the nascent Industrial Revolution.

After the Little Ice Age, the glaciers retreated slowly, then quickly. The recession was interrupted briefly for thirty-nine years (1941–1979), during a brief cooling trend. Nonetheless, any temperature rebound to be expected from the emergence from the Little Ice Age happened by 1960 or so. Since 1980, temperatures have climbed, and the remaining glaciers have declined steadily. Remarkably, on a smaller scale, today's glaciers—the last holdouts—are carving the mountains still. Every day, slabs of ice inch down the face of the mountains, gouging out tomorrow's landscape. Thanks to global warming, such crafting of the mountainsides may come to an abrupt end.

Modern warming differs from the mild period between the Pleistocene and the Little Ice Age in the level of carbon dioxide in the atmosphere. Around six thousand years ago, during the interglacial, the level was between 260 and 280 parts per million—assayed from air bubbles trapped in old Antarctic ice. In the mid-1880s, with the burning of coal in factories and homes, the atmosphere was carrying a slightly greater load: firmly at 280 ppm carbon dioxide. But, by 1959, the mean had jumped to 315 ppm. Today, it's pushing 400 ppm and climbing at 2 ppm each year. That's an increase of nearly forty percent in just over a century. Like other greenhouse gases, carbon dioxide shields the planet, allowing light through but limiting the escape of heat. In this way, it acts much like windshield glass on a car left out in the sun: Heat builds up inside the automobile and is unable to flee. The United Nations Intergovernmental Panel on Climate Change (IPCC) says it is "very likely" that global warming is due to the human release of heat-trapping gases, indicating a probability of man's primary role as greater than ninety percent.

While the existence of global warming is only now finding general acceptance, the connection between industrialization and climate change was appreciated by some scientists as far back as the 1890s. Thanks to the "glasslike" quality of the gases, it was realized that burning wood and fossil fuels must, in time, produce warming. The early scientists—most notably, Swedish chemist and Nobel laureate Svante Arrhenius—believed the effect would be slow since the oceans would function as a huge sponge, soaking up the additional carbon dioxide. This safeguard has proven to be inefficient, however; the oceans cannot keep pace with the current loading of gases from cars, furnaces, smokestacks, and other sources. The oceans and the climate are simply overwhelmed.

Dan Fagre picks up a spreadsheet from his easy chair and steps across the creaky floorboards to his desk. His table, dominated by a three-screen computer, is nestled in a corner with a window at his side. The light snow is steady now, looking like rain under a strobe of white light. Peeking over the far trees, the sun reaches its noontime height, a sickle-like arc that barely crowns the trees. Even at the sun's zenith, the winter woods are dark.

Fagre flips through his inventory of the Park's glaciers. He has just tracked the thirty-seven ice fields named in 1966 to see what's left, using 2005 as the cutoff (the last survey with aerial photography). About one-third had died. Boulder Glacier, Red Eagle, Baby Glacier, and nine others were considered so small, under 25 acres (100,000 square meters) last summer, that they have been dropped from the active list. This winter brings two fresh casualties: Shepard Glacier and Miche Wabun, too small now to move downhill. The twenty-five remaining, active glaciers lost acreage, too, but for now qualify as moving fields of ice.

"It has become clear," Fagre says, "that all of our glaciers will disappear. It's just a matter of when. There is a lag time between the loading and the effect of carbon dioxide: Yesterday's exhaust and smoke have already sealed their fate." He shakes his head, not wanting to believe his own verdict.

Fagre would post his news on his Web site in April. First he needs to crunch a few more numbers. Measuring the glaciers is a continual, year-round effort. The diagnostic work at the lab will consume the winter. Even through spring, he divides his time between grinding numbers, trekking over snow, and repairing gear. His winter outings include hunting likely avalanches and measuring the depth of the snow. He always hopes for more winter snowfall to feed his glaciers. This optimism stands in contrast to the reality of the waning ice. But, as I discover, he is always betting with the glaciers. Dan Fagre is the patron saint of lost causes.

His tools of the trade for field endeavors have changed over time. In 1966, glaciers were measured with aerial photography. This was the method still in force when Fagre arrived in Glacier in 1991. At that time, GPS equipment was either in the hands of the military or too bulky and heavy for fieldwork. In submitting his first budgets, Fagre planned to use survey equipment—sextants and other gear employed at construction sites—to monitor the recession of glaciers. But by the time his money came through, GPS gear had been miniaturized enough to allow carrying it in the field. Since then, every other year or so, satellite positioning equipment has evolved, becoming more compact. Today, Fagre uses GPS gear, ice radar, and lasers to measure glacial attrition.

But when Fagre was hired by the government to survey the Park's glaciers, he was short on technology. Meanwhile, the aerial photographs, far removed from eye-level data from the field, seemed abstract. So he decided to get the lay of the land. He hiked to Grinnell and Sperry, the two most accessible glaciers. And, indulging his passion for mountaineering, he made plans for a technical climb of Blackfoot Mountain and its namesake glacier—pure adventure, no science. While not a first ascent, the proposed climb was a watershed moment for Fagre—he would add a Montana peak to his quiver.

Blackfoot Mountain (9,597 feet; 2,925 meters) is one of five large peaks that surround Gunsight Basin, a north-facing bowl that is home to several glaciers, the highest concentration of ice in the Park. At one time, all these ice fields were joined in a single mass. As recently as the 1930s, the two largest progeny, Blackfoot and Jackson Glaciers, were one and the same. Melting subsequently divided them into two, like one of those plastic broken hearts: two lobes descending to a point. In September 1992, when Fagre and a team of five approached Blackfoot Glacier for the ascent, this secondary ice field had become the largest in the Park, over five hundred acres (two square kilometers). It is now ranked number two, right behind Harrison. Melting in the Park is uneven: Some glaciers wane faster, others more slowly.

Henry L. Stimson, an early Glacier explorer and later Secretary of State, was the first to climb Blackfoot Mountain in 1892. He named it for the Native American tribe that claimed the territory and which ironically would be paid a pittance to turn over the land to the federal government. Stimson was admired by other early pilgrims and publicists, such as George Bird Grinnell, publisher of *Field and Stream*, for his courage in the mountains. In the backcountry at night, legend has it, Stimson would hunt bear with bravado. He'd sit right next to grizzly bait with a single-load rifle at his side. The speed of a grizzly precluded reloading for a second shot. He had one chance with the rifle, or else.

For the centennial ascent of Blackfoot, the climbing team comprised employees of the U.S. Department of Interior, mother agency to both the National Park Service and the USGS. They hiked with ski poles, southwest in single file across six miles of forest and meadows to Gunsight Pass, where the rock and ice begins, avoiding the bears—speaking loudly and carrying big sticks.

At Gunsight Lake, a large tarn at the western edge of Jackson Glacier, they dug in for the night. The stars stood out like lanterns

against a black sea, assuring a blue and cold morning. At daybreak, the team made an abrupt left-hand turn, now with a heading southeast toward Blackfoot. A faint trail brought them to the
lower edge of Jackson Glacier's lateral moraine. Passing around
the moraine, Fagre and his group came upon the first obstacle of
the trip: a roaring torrent pouring out of Jackson Glacier. About
two hundred feet upstream, a huge boulder sat at midstream, six
feet from either bank. One by one, the men leapt to the boulder
and, without pausing, barreled over the second stage to the opposite shore. Nobody got wet.

In another hour, they reached a black waterfall that drained
the western lobe of Blackfoot Glacier. Just above the cascade, they
bypassed a snowfield by hiking along a low rocky ridge, which at
its upper end projected onto the lower lip of the glacier. When
Dan Fagre quickly stepped over the lip of Blackfoot, he was overcome by the vista, the scale of the ice. He thought to himself,
"What a glacier! Might as well be Alaska." The ice field stretched
nearly a mile, unbroken except for prominent crevasses, to the
pyramidal summit ridge and the peak beyond. The first enormous crevasse, more than a hundred feet long and forty feet wide,
yawned at him. "You could drop a Greyhound bus in there," he
said to a teammate. The crevasse's breadth befitted the size of
the glacier—a mountain of ice.

Donning ice axes and crampons, the men walked around the
gaping hole, a detour which set them behind ten minutes, the first
of many such tangents. Up ahead, the crevasse field was marked
with random tears, like the strips of a shredded white bedsheet.
Some of the gaps would be surmountable by a bold jump, but others would require roundabout navigation. Crevasses are one of

the hazards of glacier travel. If a climber falls into one, unroped and alone, it is nearly impossible to extract him. The victim slides rapidly down the slick casement and, with the force of the fall, becomes wedged between the converging walls. Suffocation and freezing often compete as a cause of death. In early summer, glaciers are typically covered with snow but, here, late in the season, the crevasses were exposed, visible, and easy to avoid. Still, Fagre's team was roped.

From the toe of the glacier, Fagre scanned the summit of Blackfoot Mountain, 2,500 feet (762 meters) above him. Clouds encircled the peak. The route was not a straight line but a zigzag of rock and ice that would take some shrewd navigating. The sedimentary rock of Glacier's peaks has fractured into a series of gullies and ledges and brittle walls. Climbing a summit like Blackfoot requires negotiating these obstacles with care. It's wise to avoid standing directly beneath another climber due to the hazard of rockfall. And pitons don't always stay in their cracks. The rock shatters easily, which can harry a mountaineer. These hazards are compounded when the rock is coated with ice.

Looming above Fagre, a sixty-foot ice cliff—a pale blue tower—rose above a steep rock wall. A jumble of broken ice in two sections sat like bookends at either side of the cliff, preventing a lateral assault. Farther west, they found another rock wall with a snow ramp. Here was a safer route. With a belay, the leader ascended the snow onto the rock wall. It was tricky to maneuver across the rock with crampons; the metal tips could trip up members of the team. So they skirted onto the ice. But every form of refuge has its price. Fagre and his mates had to "front point," kicking the forward chisels of their crampons into the "verglas," the thin coating of ice on rock. This was "rather unnerving," Fagre said to me later. Just in the nick of time, they came upon a ledge and a welcome rest.

Back on the glacier proper, they had a direct ascent up snow and ice fields to the summit ridge. The slope steepened near the top, and each man was belayed over the crest of the ridge. Like all tough ascents, the climb left the men in awe of the mountain, but most humbling was what they found on the summit. A mountain goat trail traversed the peak, right to the pinnacle. Goats had regularly trespassed on the summit more easily than each of the men.

The sun came out for the team's brief celebration on top. The view at all points of the compass was the main reward for the day. On the far side of the summit, Pumpelly Glacier clung to the 9,000-foot (2,743-meter) side of the mountain, an impressive ice field just larger than Jackson—each over 250 acres (1 square kilometer). Over its prow, Mt. Stimson, at 10,142 feet (3,091 meters), dominated the southern expanse. To the west, Mt. Jackson (10,052 feet; 3,064 meters) obscured Lake McDonald and the deep gouge left by the McDonald Glacier in the Pleistocene. This gorge and other valleys were overshadowed by mountains looming more than a mile above. To the north, peaks extended past Mt. Gould and Grinnell Point to the Canadian border and beyond. Like jagged teeth, these summits bit the sky in all directions, an unbroken range. The panorama was impossible to capture except on eight frames of film.

F. E. Matthes, a Blackfoot Mountain summiter in 1904, wrote: "Here is a scene which dwarfs Yosemite Valley and makes the Grand Canyon seem commonplace." Hyperbole to be sure, but there's no mistaking his awe.

Fagre took away another piece of the mountain: the image of

ice slipping away from the rock. In another eighteen years, the ice cliff that had challenged their ascent would have disappeared, melted downstream and gone out to sea. The center cannot hold. Even that day, it was coming loose from its moorings.

Fagre grew concerned that the largest glacier in the Park was melting fast, perhaps swifter than many others. He vowed to return to take its pulse with whatever resources he could muster. So, after the climb, he settled into the cabin to brainstorm. He analyzed aerial photographs of Blackfoot and other glaciers. He assembled his research team—an assistant and a couple of graduate students—and ordered equipment. He also pored over an obscure science article by Paul Carrara and R. G. McGimsey.

The 1988 publication, essentially a map showing terminal moraines (the far reach of ice at its maximum) as well as the current reach of glaciers in the Gunsight Basin, was a gold mine. The map charted positions of Blackfoot and Jackson Glaciers for various years between 1850 and 1979 with the benefit of glacial moraine dating, tree-ring analysis, field notes, and aerial photography. Fagre recognized the value of the resource immediately. Historically, it showed the profile of glaciers going back in time to the moment the Little Ice Age ended. Could the data be projected into the future?

At about this time, Fagre ran into Myrna and Charles Hall, two ecological modelers, a husband-and-wife team, at a conference and got to talking. Charles had recently modeled vegetation growth in Puerto Rico using computer simulation. Myrna was looking for a master's thesis topic. Fagre had the Blackfoot data. Another marriage was at hand. The Halls believed constructing a model of the glacial movement—either waxing or waning—was feasible.

Fagre's vision of the computer simulation was not as a formal forecasting tool but as a communication device. He figured Charles's software could be adapted for projecting glacier movement in 3-D, essentially in time-lapse, on his USGS Web site. In this fashion, the model could educate the public about glacier loss. Myrna would be his partner for shaping the model, from clay to kiln.

Fagre and Hall chose the Gunsight Basin, encompassing Blackfoot and Jackson Glaciers, since it was the largest ice complex in the Park. This cirque wasn't just an ice field but a full mountain ecosystem with vegetation around the periphery. Part of the modeling would include a record of vegetation advance as the glaciers retreated. But the overwhelming draw of Blackfoot/Jackson itself was the prodigious field data that Carrara and McGimsey had gathered for that particular glacier complex, showing how it had moved. Over 129 years, the advance and retreat of ice was documented like a tailor's measurements for enlarging, shortening, and hemming a suit of clothes.

In the 1980s, Paul Carrara found a very different environment in the Park compared to its inception in 1910. Glacier National Park had lost most of its namesake glaciers. More than two-thirds of the estimated 150 glaciers dominating the area in 1850 had vanished. The combined surviving ice mass was greatly reduced in size. The Gunsight Basin housed twenty-seven glaciers and over 5,190 acres (21 square kilometers) of ice in 1850. By 1979, its remaining ten glaciers held one-third the amount of ice. During the lifespan of the Park, from 1910 to 1980, the local mean summer temperature had increased 2.99 degrees F (1.66 degrees C). These events mirrored a global pattern of glacial retreat and climate change that, in aggregate, was viewed as an early signal of global warming. Fagre's and Hall's model would bring the advisory up to date.

From the beginning, their objective was to look at the history and possible future for a dual glacier—Blackfoot/Jackson, which was once a single ice field. It happened to be the largest glacier complex in the region, however, so its likely disappearance, with the most to lose, might coincide with a total loss of ice in the Park. In other words, if Blackfoot/Jackson vanished, all the other glaciers would likely have gone before. Recognizing this potential implication, they set their sights high. They planned to develop a spatial model that incorporated the most accurate records of the main causes of glacial advance and retreat—namely temperature and precipitation—over thirteen decades. They would plot the historic glacier margins and corresponding climate parameters and then set up modeling for two possible future scenarios, over the period 1980–2100.

The legwork in the 1990s was split between West Glacier and SUNY at Syracuse, where Myrna Hall was a graduate student. First on the agenda was to digitize the glacier extents from Carrara's historic maps using a GIS (Global Information System) software package. This transference took some time since it comprised terminal moraine positions in a twenty-five square mile (sixty-five square kilometer) area. Also, temperature records for each decade were obtained from three weather stations circling the Park. All these data were incorporated. And one more parameter was needed: Since glacial advance and retreat are largely a balance between winter snow precipitation and summer melting temperatures, it was important to include precipitation in the model, as well. Out of a variety of candidate forces of glacial movement, the model would confirm that the strongest predictors of glacier area historically were July/August temperatures and mean annual precipitation, both snow and rain.

Now, back to the principals: Myrna Hall wrote a FORTRAN computer simulation model, called GLACPRED, that would predict glacial advance or retreat for each decade from 1990 to 2100 based on melting rates from historical data. The team fed two possible climate scenarios—that is, two alternative futures—into the simulator. The first, called the carbon-doubling scenario, was based on EPA's 1989 forecast of a doubling of atmospheric carbon dioxide by year 2030. This prediction held a concurrent worldwide temperature response of 4.5 degrees F (2.5 degrees C) by 2050. Tied to this first paradigm was a five to ten percent increase in winter precipitation. (Precipitation increases with global warming because the heat stimulates more evaporation from the world's oceans and because warm air holds more moisture than cold air does.) The second scenario, entitled the linear temperature projection, assumed a variable climate not linked to anthropogenic increases in carbon dioxide. This reflected a more gradual (and natural) emergence from the Little Ice Age. In this world, temperature would rise by only 0.4 degrees F (0.22 degrees C) by 2050.

Even before the model was fully up and running, Fagre and Hall posed a couple of hypotheses about the two future climate scenarios. They were anxious to get a reading on their theories. Since a warming planet would bring on additional snow in the first scenario, they felt melting would likely be buffered by the accumulation of ice, thus slowing any losses to the glacier. They imagined the ice surviving past the middle of the twenty-first century. On the other hand, for the conservative linear temperature scenario, in which there would be no precipitation increase, they thought glaciers would melt at the rate of the 1980s and vanish earlier, by 2050.

Fagre and Hall were wrong on both counts. The model had its own logic, and the results were a surprise. With a doubling of carbon dioxide, and a concomitant increase in snow, no buffering actually occurred. The elevated summer temperatures simply overwhelmed the system and the additional snow. Blackfoot/Jackson Glacier and its neighbors melted rapidly as a result, losing about 370 acres (1.5 square kilometers) per decade. By 2030, the glaciers—the largest in the Park—were completely gone. Meanwhile, in the second (linear temperature) scenario, melting was *slower* than expected—approximately 52 acres (0.21 square kilometers) per decade. On that stage, the glaciers in the field lasted beyond the year 2100. But the glaciers in the field were already disappearing at a faster pace than predicted by this linear decline. Clearly, more than normal attrition was at work.

Yet here's the real kicker: In the first scenario, with a doubling of carbon dioxide loading, the glaciers would have a lifespan of thirty years, rather than sixty or more. Their timeline was cut in half.

Fagre's and Hall's hypotheses had been too conservative on all fronts.

On his Web site, Fagre projected the 3-D simulation of the Gunsight Basin model, including Blackfoot and Jackson Glaciers, by draping each ten-year image from 1850 to 2100 over the terrain. This produces a time-lapse visualization of the changing landscape in twenty-six snapshots over 250 years. On screen, the model resembles a topographic relief map, accented by a palette of colors that represent natural features ranging from rock and ice to alpine trees and shrubs. A spiderweb of black contour lines defines the peaks and ridges and valleys from the Gunsight Lake Basin to the top of Blackfoot Mountain and Mt. Jackson. At the start of the simulation, in 1850, there is one large glacier; by 1940, there are over a dozen offspring. The time lapse refreshes every second or so—that's ten or more years each second. The effect is nearly a video-speed progression from 1850 to 2100. It takes about thirty seconds to go from a blanket of white, before the Civil War, to a post-modern, gray rockscape. The future holds no ice, only advancing grasslands and trees—for glacier watchers, a barren landscape.

Two of the most striking results of the carbon-doubling scenario were the rapid elevation of summer temperatures and the corresponding percentage of melting ice. By 2100, local mean temperatures were expected to reach 67.6 degrees F (19.8 degrees C), an increase of 5.9 degrees F (3.3 degrees C). This exceeds the warnings of the IPCC for global mean temperature in 2050. The model crunched these numbers and made more forecasts: From a baseline ice coverage in Gunsight Basin of 1,535 acres (6.21 square kilometers) in 1980, including Blackfoot/Jackson, ice was reduced sixty percent by 2010 and ninety percent by 2020, before disappearing altogether by 2030.

The model, which is still displayed on the Web site, also mirrors what researchers have uncovered in the historical record. Over the past hundred years in Glacier National Park, precipitation has increased by ten to fourteen percent, depending on how it's measured. Much of that water fell as snow. If temperatures had been constant, the glaciers would be growing. But they're not; they're melting. This evidence is clear proof of climate warming and its impact locally in Montana.

The response to the model and its projections upon publication in 2003 was swift. The scientific community embraced the findings; climate skeptics didn't weigh in. Al Gore voiced his support. Even more immediate was the reaction of the media. Nearly every news service around the world picked up the story. To their surprise, Fagre and Hall became darlings of the national and global press. Two elements made the story newsworthy. The fact that a famous national park was expected to lose the resource behind its name was obviously a draw. The second news was the timeline—the date of the demise. But while the model had projected the melting of several glaciers, it had not addressed the fate of all twenty-seven glaciers in the Park. The news agencies made that extrapolation. If the largest glacier could melt by 2030, then wouldn't all the other ice fields have vanished before it? "A logical conclusion," says Fagre, "but we didn't make that proclamation." Still, the tragic headline appeared again and again: GLACIER NATIONAL PARK TO LOSE ALL ITS GLACIERS IN 30 YEARS.

Fagre has another criticism, this one reserved for himself. By the time the model was launched in 2003, the computer software was over ten years old. "We had various delays," he says. "It took a long time—ten years—to publish our findings. Looking back on it now, our simulation seems primitive. By today's standards, our techniques were crude." Surprisingly, though, the model predicted the future trend with remarkable accuracy—just a little modest on the numbers.

At the cabin, during that winter day of light snow, Fagre talks about updating the model. To his mind, the precision of the forecast is in question. The perfectionist back at work. He has another decade of weather data now, which shows warmer temperatures than they had postulated. From visual inspection in 2009, he suspects Blackfoot/Jackson has shrunk beyond the model's 2010 boundaries. He would need to visit Blackfoot again to measure the current margins with GPS equipment to make sure. Then he could determine if a revised climate model would predict the acreage of today's glaciers accurately. Even at their accelerated pace.

A scientist is always curious to see if he is right in hindsight. But the double-checking would have to wait for the summer season. Right now, Fagre must complete the ice inventory for Glacier National Park's centennial: May 11, 2010. His deadline for the comprehensive profile is fast approaching. The USGS headquarters in Reston must review the figures first before they are deemed official. He rummages around his office, searching for some more data—the area calculations for all twenty-five remaining glaciers in the Park—as I listen, still on speakerphone.

"Now that it's nearly 2010, the 2005 data will be five years out of date," he says, "but it's the last year for which we have aerial photographs for the glaciers that we've been unable to reach on foot. In any event, it'll show a forty-year trend."

While Fagre collects and crunches some more numbers, the snow stops outside. He pours one more cup of coffee, then sets it aside with regret: It is too late in the morning for caffeine, even for a coffee aficionado. The wind is quiet, enveloping the small cabin in a white silence. The late-morning clouds break into shreds, revealing a canopy of blue. With fair skies overhead, Fagre decides he will cross-country ski home at the end of the day, his preferred mode of travel.

Before signing off, I ask Fagre if the government approval process will take long.

"It shouldn't," he says. "Aerial photographs don't lie. The ice is vanishing right under our noses."

. . .

May eleventh: When the centennial celebration in West Glacier gets underway, the senior officials stand on stage, reveling in the moment. The government leaders are proud of their century of achievement. But they don't make any announcement of Fagre's findings. Little mention of the shrinking ice fields. No mention of the recent loss of Shepard and Miche Wabun. The breakthrough inventory and report will be relegated to a back page of the CCME Web site. Instead, congratulations are passed around for the stewardship of the nation's fifth largest national park. Another two million visitors will be expected that summer. Praise is bestowed upon the Park's beauty. "Glacier connects us to the very core of our nature," says park superintendent Chas Cartwright. He is eloquent about this connection and about humanity's role as steward. As custodians of a vast wilderness, the NPS has done well, his logic goes, and the nation can look forward to another hundred years.

Without glaciers, of course.

The Falcon and the Falconer

The last obstacle in the ten-mile climb to Sperry Glacier—a 4,800-foot (1,463-meter) elevation gain—is a rock stairway cut in the raw stone, called "The Staircase." The ordinary name underplays the severity of the ascent. It has the exposure and verticality of a fire escape on a New York skyscraper. The steps are set in a triangular corner, which reaches up to a lofty window in the last 200-foot-high bench below the glacier. Mercifully, a cable railing is anchored to the rock. My right hand clenches it until knuckles rub raw. But there's no alternative. It's the only way to the window at the top.

The Staircase is precarious in the best of conditions, but rain or snow makes the limestone slick and treacherous. Today, nature's cruel joke is the wind. A forty-mile-per-hour burst of air tears through the vertical chamber, seemingly from all sides, as if the skin were being flayed off the building. The rock wall must be ricocheting the breeze: It's hard to make out its origin. The wind makes me more careful. Every step—there are thirty to my count—requires hefting my foot cautiously and, with one hand on the cable and one hand on my trekking pole, rolling my gait upward into the dark stairwell. Halfway up, my legs feel as heavy as waterlogged timber. But each new step brings me closer to heaven: Sunshine bathes the top of the chute.

At the 8,000-foot (2,438-meter) summit, my comrades and I stand up and cheer. The loudest hoots and hollers come from our guides, Stacey Bengtson (Lindsey's aunt) and Mike Davies. The other climbers are from Colorado. I stayed with them the night before at Sperry Chalet, a stone-and-wood lodge perched a couple of thousand feet below. We are booked there for the duration of the trip—three days in mid-August 2011, after a wet winter. Dan Fagre and his team, on a faster circuit of two days up and back, would come shortly and stay in a tent. They were used to roughing it up here. Sperry is their benchmark glacier, the one they monitor without fail every year.

We traverse three snowfields to reach the glacier. Without my eyes trained on my feet, I slip and slide across the snow, all the time scanning the horizon for peaks that I may know. I dig through my pack for a map. But charts never did mountains any justice. The sheer faces must be seen to identify them. A ring of summits surrounds us, dense as whitecaps on the sea. I raise my binoculars. The broad buttress of Reynolds Mountain stands in the distance like an open baseball mitt. Also recognizable (from all directions) is Going-to-the-Sun Mountain, with its pyramidal summit. The back side of Mt. Gould rises like a tower above the Garden Wall. Cannon Mountain must be the one in the foreground. Over thirty peaks touch the sky, most of their names unknown to me. I'll keep it like that, refolding the map. It's hard to connect the dots. Besides, a little mystery befits the realm. I celebrate its wildness.

On the third rock rib (past the third snowfield), I scan the glacier with my binoculars again, taking in the expanse of ice. Sperry Glacier is facing north—its saving grace. Its columns of ice evade the sunlight, melting less. Icicles dangle from the rock faces like a white beard. I quickly make comparisons to Grinnell Glacier, which I've now visited twice. Sperry boasts a steep headwall, laced with horizontal crevasses, as does Grinnell, but there the similarity ends. While Grinnell sits in a perfect bowl with a symmetrical layout of ice, Sperry is lopsided, tilted to the right. (Right-handed, that is, when looking downslope, as I am now.) It is also bigger. The ice would fill thirty-three city blocks; it is over 833,000 square meters broad—about 206 acres. Grinnell is three-quarters as large. The glacier runs from just below the east summit of Gunsight Mountain (9,758 feet; 2,974 meters), which resembles a turret behind it, to halfway across the shallow cirque. And it hangs high: Right to left, the glacier straddles the lofty ground between two adjacent peaks. It is perched aloft on the bench between, as if on the balcony of a fortress.

Only the fortress has been breached. The ice wall at the terminus used to be a dramatic 500 feet (152 meters) tall, the prow looming like a huge embankment. Now the terminus simply peters out. You can walk right up the rampart. And the ramp starts farther uphill. Today, the rock is exposed midway up the cirque as if a tide were receding. The river of ice descends toward the moraines, to be sure, but stops short. At its farthest extent (in 1850), the glacier left its impression, however: a unique, long hill of debris. This moraine suggests a battle mound and is the only relief at the bottom.

Between the glacier and moraine is meltwater. The ice terminates in several streams that feed three small glacial ponds, before rejoining to pour over a precipice. That enormous cliff, over 3,000 feet (914 meters) high, harbors several dramatic waterfalls. One of these cataracts contributes Sperry meltwater to Avalanche Lake far below. All these attributes make Sperry a rarity. In fact, the ice field's perch over cliff and cascade gives Sperry a special designation: It's called a "hanging glacier." From below, the glacier is hidden in its balcony, draping anonymously and precariously above the rock wall.

Sperry Glacier was photographed first in 1887 and mapped in 1901. The first scientific profiles and measurements were published in 1914. At that time, Sperry still occupied 843 acres (3.41 square kilometers), roughly ninety percent of its maximum extent—that is, the area up to the moraines—at the end of the Little Ice Age. Over the next decade or so, the retreat accelerated, and the glacier split into one central panel and two smaller wings. By 1927, the main ice mass covered only 492 acres (1.99 square kilometers). The area in retreat was matched by thinning of the glacier's mass throughout the twentieth century. One researcher calculated a mean annual height loss of nearly 3 meters (9.8 feet) between 1938 and 1946 on the glacier proper, and a ninety-meter elevation loss from 1913 to 1946 at the glacier's terminus. That's more than 295 vertical feet of ice removed, the height of a twenty-four story building. Essentially, the ice wall was shaved in half. In the next twenty years that were studied intently (1949–1969), the foot of the glacier thinned by another 115 feet (35 meters). That's the last of the thickness measurements made at mid-century. But Sperry's area reduction for 1966–2005 has been documented with aerial photography: The main glacier declined from 331 acres (1.34 square kilometers) to 216 acres (0.87 square kilometers), a drop of thirty-five percent. Those forty years saw some of the hottest summers on record. The two ice wings completely disappeared.

Today, a Sunday, most of the glacier is still covered with snow like swirls of icing on a cake. This is unusual so late in the season, but the winter snowpack nearly topped the 1969 record and is slow to melt. Nonetheless, here and there, patches of ice reflect the sunlight. Several crevasses are also exposed, especially at the headwall, smiling with expectation. They are bound to grow. As the snow melts, more will appear. Even in the belly of the glacier, cracks and fissures are breaking through. One wonders how ubiquitous the network of crevasses might be. The overlying snow gives a false security.

Dan Fagre and his team of two—Kevin Jacks, forty, and Erich Peitzsch, thirty-three—arrive two days later with impressive packs of gear, impressive not so much for their bulk as for the lethal accessories tied on top: ice axes, crampons, ice screws, and ropes. It is a tight squeeze through the Staircase—like threading the eye of a needle. A snag would be dangerous, but the three men slip through and break free cleanly. The wind has modified by now. Dan and his crew would give me a full report at the end of the trip.

Upon reaching the rock outpost, Fagre removes his sunglasses and stares. The view is captivating; it is hard to look away. Mountains upon mountains are snow-capped and bright. They stand out only slightly against the dappled sky, ivory pillows on a bed of the same color. Snowfields and clouds seem interchangeable. Almost impressionistic. Fagre says, "I sure would like to paint this scene someday," indulging his boyhood passion for landscapes in oil.

"Why don't you?" someone says.

"While I'm resting—after a week of glaciers, all I seem to have time for is painting my house."

Dropping their heavy packs, the men rearrange their equipment, getting set to make their rounds. From the beginning, field notebooks at the ready, it is clear the diagnostic work will be nothing like that of Grinnell. Here, Fagre must listen to the glacier more closely.

Three years before, Grinnell Glacier had been sixty-six percent free of snow, allowing measurements of the exposed ice area; this year Sperry is nearly covered. The snow hides the ice like a white shawl. Overall area measurements with GPS will be impossible. But the day is not lost. The team agrees a different technique for surveying the glacier is required. And they're in luck. Fagre has one in his hip pocket, a trick for bypassing area and measuring the mass of the ice instead. The mass test will be an even better diagnostic tool. He's been refining the procedure for six years.

But first: precautions. To combat the threat of hidden crevasses, the three men don climbing harnesses and tie up to a fifty-meter, blue-and-green perlon rope. They attach crampons to their boots and arm themselves with ice axes. In case one of them falls into a crevasse, each man has ascenders to climb the rope back up to his comrades. The scientists will navigate the snow with half a rope length between each man. It will be critical that the line is taut. Tension on the line will detect a fall quickly—like a quiver. If there is a fall, the second man will drive his axe into the ice deck, and the lower man will climb up the rope and out. That's if everything goes according to plan, but mountains sometimes have schemes of their own.

Before tackling the ice, the men are curious to check on their weather station, a remote listening post running since 2006. It sits on the prow of the rock rib just above the men, tethered to the rock. Resembling a lunar module, though even more compact, the station operates with only minor maintenance as long as the snow doesn't cover its solar panel. Snowfall can exceed thirty feet at Sperry. Thus, it shuts down each winter until late spring. But all summer long it senses and stores data on temperature, relative humidity, solar radiation, and wind. The solar panel recharges a car battery that runs the whole thing. Fagre says, "We put the station here, carrying the works on our backs, to see if any weather conditions were different at this glacier. In a few more years, we'll have enough data to make a judgment. Maybe it's colder here than at similar places in the Park where there's no ice. Maybe it gets less sun. That would help explain why Sperry exists."

The men gather around the weather station, with their ropes still in coils over their shoulders. Their crampons clink against the rock. The wind gauge is whirring, and the weathervane is darting back and forth. Kevin drops to one knee, setting his ice axe down gently, to examine one of the guy wires to the station. It's loose. The wind is a continual threat to the equipment. Being covered by up to thirty feet of snow in winter doesn't phase the station, but wind could damage the works.

"Dan, I'll tighten this guy wire at the end of the day," says Kevin.

"Yeah, we should check all the sensors, too," says Fagre. "Looks like the battery is running."

"Good. I'm in no mood to lug another car battery up here."

"That's how you get a post with Dan," Erich says. "He sees how many car batteries you can carry, then maybe you get the job."

There are six USGS weather stations in all—all in remote alpine settings, though the ten-mile trudge to Sperry holds the record for inaccessibility. While the others transmit weather data back to the lab (and do it all year long), the seasonal Sperry information must be downloaded on site. Kevin carries a flash memory device to do it.

Kevin and Erich share similar early backgrounds, though their training took different turns along the way. Both left small towns and the flatlands of the East Coast for the heights and deep snow of the Cascades and the Sierras. From there, interests diverged. Kevin spent his time on various ski patrols, preferably racing downhill; Erich headed uphill—he became a climber. Kevin delved into ski mountaineering on the Columbia Icefield and in the Selkirks of Canada, still favoring the descent. Erich honed his climbing and mountaineering skills as an instructor at Colorado Outward Bound. Later he earned his master's in snow science from Montana State University. Meanwhile, Kevin puts his geology degree from the State University of New York (SUNY) at Plattsburgh to use as a field technician for Dan Fagre. They have come full circle from their small boyhood towns: Their skills complement each other on the glacier. One keeps you safe on the ascent, the other gets you out of trouble fast at the end of the day.

Kevin picks up his ice axe. He joins Fagre and Erich at the edge of the rock rib where it meets the ice and snow. They uncoil the rope and tie bowlines to clip into their harnesses. The two men look to Fagre for instructions.

Dan Fagre just points, unusual for a loquacious man. Ahead, about two hundred yards into the glacier, toward the headwall, is a bald spot. The snow has melted away or been scoured by the wind to reveal a stretch of disturbed ice and crevasses about the size of a football field. The ice is dark and shiny, like the frozen film over a flooded parking lot.

"No doubt there's a bulge of bedrock beneath it," says Fagre, "which the ice is sliding over. That's why we see fractured ice."

The ice itself is exposed because the snow is thinner all around the ice patch. "The snow has been blown away and pillowed up else-where," Fagre says. Some curves along a glacier bowl can act like tunnels of rushing wind.

Those wind spouts have scoured at least five ice patches free of snow. Loosening up now, Fagre tells the men that they need to measure each patch to add up how much ice is exposed on the glacier. They will deduct this from Sperry's overall area (as fig-ured last year) to calculate the percentage of snow cover. This is akin to measuring the net surface of a block of Swiss cheese: The holes don't count. At the end of summer, the snow coverage on a glacier is an indication of its health. Snow will convert into ice; enough of it and the glacier will expand. Remember: Anything around sixty percent coverage means the glacier is near equilib-rium, holding its own. Over sixty percent typically means it's growing. A glacier accumulates most snow—and makes most ice—in its upper reaches. Most melting happens at the bottom. For these reasons, snow cover usually drapes only the head of a glacier in late summer. This summer breaks that typical pattern: Snow runs all the way to the foot. This demonstrates the variabil-ity in mountain weather; still, on average recently, melting has outpaced the annual snowfall.

Fagre dispatches his crew toward the first ice patch. "Erich, you go first, my man," he says. Erich is the most experienced mountain-eer, having cut his teeth on the crevasse-laden slopes of Mt. Hood (11,240 feet; 3,426 meters) while attending college in Oregon.

Because of his skill, he is often the lead man on the rope. He has another advantage: He is skinny as a razor, the least likely of the three to fall in.

Once on the dark ice, Kevin belays Erich as he circumnavigates the patch. Erich records the ice's perimeter with his GPS equipment, making hundreds of readings that he'll plot back in the lab. Pacing off the ice patch takes forty minutes, but most of that is taken up by rope work. Each belay is a lifeline. Like a spoke, the rope plays out from hub to periphery. The team crisscrosses the oval ice as if spinning a web of silk.

Toward the end of the circuit, Kevin pays out the line to Erich so he can cross a snowbridge. The three-foot span connects two sides of a crevasse with the void looming underneath. The snow bridge likely formed when snowfall, adhering to the lips of the crevasse, spanned the gap. Meanwhile, an ice bridge represents an incomplete tear in the ice, as the glacier slowly moves.

Ice is fluid in a glacier and advances like brittle tar flowing downhill. The weight and pressure from the ice above makes it ooze. It slides and stretches and, when it moves faster, cracks, especially around steep corners or bumps in the road. The face of a glacier manifests whatever its rump is sitting on. Its countenance may be smooth or pitted, a gentle smile or a deep grin. The scale of the grimace is proportional to the speed of the ice and hence the steepness of the terrain underneath. When the ice slides over a hump of rock and falls precipitously down the other side, it cracks slightly; its surface looks like it has been cut into small ribbons. But when the ice falls more rapidly down a steep, broad decline, the face fills with long, deep fissures—a crevasse field. On this scale, Erich is standing over a moderate disturbance, maybe a rock outcrop or two—a steeper slope than above. But it's more than a hundred feet beneath his toes. In between is a labyrinth of ice: It's possibly riddled with holes, and he's careful not to fall in. Gingerly, he steps off the bridge onto solid ice.

An ice bridge figures largely in my own mountaineering history, and I will share the story with Dan once he returns to the lab. The year was 1974—it was August, the month Nixon resigned. I was climbing the Emmons/Winthrop Glacier Route on Mt. Rainier (14,410 feet; 4,392 meters) in the dark hours of predawn, following another team that sported headlamps. I could see their lights searching the route above. Three of us—Jake Stout, Bob Underhill, and myself—were strung on a single line below them, like today's rope of Fagre and his team. At dawn we turned our headlamps off and continued upward. A bright blue sky. We could see the crampon tracks in the ice of the three-person team we were following. The summit was a couple of thousand feet above.

Suddenly, we came upon a crevasse—blue and deadly. The tracks ahead of us went up to the edge and simply disappeared. The hole gaped six feet wide. No sign on the opposite side. We looked at each other in disbelief. Bob yelled into the abyss. No response. Jake uncoiled a second rope and descended. Once the rope was taut, Bob and I heard more yelling, echoing off the chamber. After a long delay, Jake ascended the rope with his report.

Yes, there were three climbers in the crevasse—on three ledges—about 120 feet (37 meters) down. They had been crossing an ice bridge when it collapsed under the leader's weight, sending all three—a doctor, his wife, and a friend—bouncing down the narrowing sides of the crevasse, like sand jumping around an hourglass. Their rope, their only means of escape, lay draped over their bodies like a fallen curtain. All three were alive but injured. Their mistake: They had not climbed with the rope stretched out between them. They crossed the ice bridge in too tight a formation, with the rope coiled, and were yanked one by one into the abyss.

We were their only hope for a rescue. I have often reflected, in the years since, what extraordinary timing it was that we happened upon the accident in the first hour of the emergency. They were stranded, and hypothermia was setting in. But they were still able to speak, to respond to our calls. Three of us in the sunshine, and three of them freezing in the hole: We had to act fast.

Since the numbers on each team were equal, we decided to match them, man for man. We set up a crude pulley system with our ice axes as anchors in the snow and our carabiners as pulleys. Each of us would go down in succession to retrieve a victim. I went first and found the doctor in a stupor on a shelf. He was bruised and bloody from what turned out to be minor cuts. I tied the rope to him, and Jake and Bob hauled him up. It was muscle-numbing work. Then they tossed me the rope end, and I ascended with a mechanical device, called a Jumar.

While Bob went after the next person—the leader, Doug Pengilly—I bundled the doctor in a sleeping bag to warm him up. Then I wrapped my arms and legs around him. He was shaking uncontrollably. But he had the wherewithal to instruct me on how to check his vital signs. His pulse seemed to be racing, his breathing fast and shallow. He was in mild shock. His face and arms were bruised red and purple. I administered first aid on his cuts as best I could.

As it happened, a Rainier mountaineering guide had seen the fall from the next ridge. He sent a colleague to radio news of the accident, and then he crossed some treacherous terrain alone to help us out. He arrived with real pulleys and another rope.

An alpine view in Glacier National Park shows three glaciers perched below Mount Gould: Gem *(center right)*, high on the ridge; Salamander *(far right)*; and Grinnell *(center)*, next to its lake. Celebrated both for its jagged peaks, carved by Pleistocene ice, and for its present-day glaciers, the park will likely lose its last twenty-five ice fields in the next decade. PHOTOGRAPHS COURTESY USGS EXCEPT WHERE NOTED

above: Lisa McKeon views Salamander Glacier and ice-strewn Upper Grinnell Lake through her camera, and clicks, while holding a 1938 photograph taken from the same stance—atop Mount Gould (9,553 feet; 2,912 meters). The two photographs will show how much the ice has retreated in the seventy-one intervening years.

opposite: Boulder Glacier in 1932 *(top)* shows a robust sheet of ice. (PHOTOGRAPH BY GEORGE GRANT, GNP ARCHIVES) In the repeat photograph taken by Lisa in 1998 *(bottom)*, the glacier is nearly gone. Today, Boulder Glacier holds less than 13.6 acres (55,000 square meters) of ice, well below the critical size for a functioning, moving glacier. Over 125 glaciers have disappeared from Montana, comprising a regional crisis. Glaciers and snowpack store nearly 75 percent of the drinking and irrigation water in the American West.

Boulder Glacier 19

Boulder Glaci
Lisa McKeon phot

above: Dan Fagre, ecologist and mountaineer, navigates through a crevasse field on Black-foot Glacier. Dan and his USGS team measure snowfall levels and avalanche hazards in winter and the melting of ice and other indicators of climate change in summer. Predictions of glacier longevity are made with computer simulations, utilizing this data.

below: In August, Dan shovels snow from a temporary pit to ascertain the density of snow that blankets Sperry Glacier. The results will help prove whether the glacier is waxing or waning.

above: The wind gauge on the Sperry Glacier weather station whirs while Lindsey Bengtson braces against the cold. Annual temperature readings surrounding the Park show a 2.4 degree F (1.33 degree C) increase since 1900—1.8 times the global average—and still climbing. The ice has retreated closely in step with the rising heat.

below: Lindsey paddles a kayak to get closer to the terminus of Grinnell Glacier, which she will measure with a Global Positioning System (GPS) device.

"In a world of climate change," says Dan Fagre, "there will be winners and losers." Perhaps one of the species to suffer will be the mountain goat *(above)*, since they depend on alpine tundra plants, which—as temperatures rise—are being overtaken by trees.

The grizzly bear *(below)*, whose population remains in excess of 300 in the park, is opportunistic and unlikely to diminish in numbers in a future of global warming. PHOTOGRAPH © DENISE DONOHUE

above: Local warming and associated drought render mountain slopes more susceptible to fire. With diminished flow from glaciers and snowpack, drier streams compound the problem for forests. In 2003, twenty-five fires burned 146,500 acres, or 13 percent of the park.

below: Forest fires are just one of many growing costs of climate change. Water rationing and avalanche danger are two other price tags. Last year, over $1 billion was spent to fight wildfires in the United States.

Icebergs and ice floes on Upper Grinnell Lake are the legacy of Grinnell Glacier melting above it. During the forty years from 1966 to 2005, the ice field's area decreased by 40 percent. By 2009, it encompassed only 143 acres (0.58 square kilometers)—a living glacier, yet scientists predict it will vanish sometime around 2020. The loss of ice in Glacier National Park foreshadows the likely extinction of thousands of glaciers in mountain ranges around the world as temperatures climb. PHOTOGRAPH © WILDNERDPIX

Quickly, he set up a professional rescue system, and we pulled up the other two easily.

By now, the doctor, who introduced himself in a quavering voice as Alan Peterson, an internist, was calming down but couldn't walk well. He seemed disoriented. To check whether he had a concussion, I asked him what the date was and where he lived. He could answer those routine questions but had no idea who the President of the United States was. I pressed him but he claimed there wasn't one. Strange, yet he was insistent. I grew concerned. But it was I who was in the dark: Ford had not yet replaced Nixon. On the mountain, I was not up on the latest national news. So I thought Peterson had had a memory lapse from the fall. Turned out he wasn't that badly injured, after all. His wife, Mary, was definitely dazed, however, and had suffered a broken collarbone. Doug Pengilly was shaken but just bruised; he walked out with us, down the mountain. The Petersons awaited a helicopter.

Before we departed the crevasse, Dr. Peterson asked me to come near. He whispered, "The first million I make is yours."

I've been calculating physicians' salaries ever since.

That's thirty-eight years of arithmetic. My climbing partners and I were just eighteen years old at the time.

Safe to say, we learned a valuable lesson: Follow the rules on a glacier. She will not forgive a mistake.

Securely on firm snow now, Erich and Kevin walk in step to the next ice patch, about one-quarter mile away. On two occasions, they must backtrack when snow bridges look dubious. Fagre trails, plotting in his mind how to survey the remaining four patches in record time. When they're finished, all five spots are measured in just over two hours—exceeding Fagre's time budget by half an hour or so.

"Pacing off the holes in the snow cover took longer than gauging last year's snow line," says Fagre. "Tracking a snow line, halfway down the glacier, is like tracing the ruffles on a dress. One long wavy hemline—but pretty simple to track."

And quick.

Today is like sewing buttons on a dress: There are several ice circles to complete. Time consuming. But he's got some flexibility in the schedule, as long as he gets off the glacier before dark. The GPS data, plotted back at the lab, will reveal that the snow covers ninety-seven percent of the glacier's area. Sperry's new acreage figure (206 acres; 0.833 square kilometers) comes from last year's ice margins, measured when they were fully exposed. That year the snow line coverage fell under the sixty percent threshold, and the glacier was losing ice. Now, for two-thirds of the year, Sperry is growing—gaining ice so far. Next month or next season could bring more recession.

But for now, for the team, it's time to celebrate. "Yes," Fagre shouts, "it's a great summer for Sperry! Right now, the scales are tipped in its favor. It may be a good season for the whole lineup, too."

Each glacier's history is determined by either snow accumulation or melting; one takes precedence. This past year, Sperry's story was all about snow. The winter of 2010–11 produced the biggest snowfall in northwest Montana in forty-two years. The regional snowpack was nearly 180 percent of the multidecadal average for the Flathead River Basin, which drains much of the west side of the Park. Fagre estimates that over forty feet of snow fell at Sperry Glacier, settling into twenty-five feet of snowpack. That's a lot of snow crystal—to melt or transform into ice.

The record snow coincides with a La Niña event, which dominated the past year and brought cold, wet weather to the Pacific Northwest and inland. La Niña is the flip side of El Niño; together they make up a cycle of Pacific Ocean temperatures and corresponding weather in the western U.S. While the warm waters of El Niño tend to reverse local weather patterns, making the Southwest cool and wet, for example, La Niña does the reverse, rendering the Southwest even warmer and drier. Farther north, La Niña can augment the heavy Montana winter with even heavier, record snows. Meanwhile, Seattle enjoys its legendary rain.

The underlying forces behind the El Niño–La Niña cycle (what is known to climatologists as the El Niño Southern Oscillation or ENSO) are complex. ENSO's best known signature is the temperature of the South Pacific from the International Date Line to the South American coast—either unusually warm (El Niño) or cool (La Niña). Since the sun's heat is stored in various currents, ocean temperatures drive the world's weather. In the case of ENSO, the atmosphere reacts in surprising ways. In La Niña years, the trade winds, which blow east to west across the equator, grow stronger; in El Niño years, they weaken. Each phase can last anywhere from a few months to a few years. Not only wind speed is affected; the direction of the wind as well as atmospheric pressure and temperature are modified. In many places, like Montana, the ENSO effects are fairly predictable. Fagre knows what kind of winter he'll have—and how much snow his glaciers will likely get—from reading the sea temperatures off Peru.

Just when ENSO starts to make sense, there is another wrench in the works: At least one other major ocean fluctuation—the Pacific Decadal Oscillation (PDO)—affects temperature and precipitation patterns in western North America over a broader span of time. The warm and cool regimes oscillate, as in El Niño, but

PDO events persist for two or three decades. The PDO was responsible for the heyday of the Monterey sardine fishery, when ocean temperatures went into a warm phase in 1925 and lasted over twenty years. (Sardines prefer warmer water.) The same warm, dry phase likely contributed to the drought of the Dust Bowl in the 1930s. (The Atlantic Multidecadal Oscillation was also positive that decade and likely influential.) Today, the PDO is entering or is on the verge of entering a cool phase. If so, whenever it is in sync with a La Niña year, it will likely compound the cooling effect, bringing more snow to Montana. But here's another wrench: Global warming could partially cancel the effect.

"We are entering or not entering a cool phase of the PDO," says Fagre, "depending on who you talk to. Climate change may be damping the cycles. It's possible. But there is no consensus yet as to whether climate change will weaken or exacerbate the PDO signals."

Glaciers respond to PDO phases, but global warming may have a longer-term effect on the ice. PDO is measured in decades while climate change will likely be measured in centuries. Consequently, Montana's glaciers are not only slipping away; they're not likely to come back anytime soon.

Yet it's tough to kill a glacier, for a glacier will grow if it's given more than half a chance. Given enough snow, Fagre says, he could build a glacier in a Wal-Mart parking lot. A glacier is hungry for snow. It consumes anywhere from two to ten feet or more to make one foot of ice. Here, in mid-August, Sperry is satiated, gobbling up the snow, a turnaround for a glacier that has seen major melt years over the past two decades, times when Sperry starved.

Those good years and bad years would even out if Sperry were stable—neither growing nor retreating over time. But, aside from minor blips like now, the record shows a steady retreat. The glacier is emaciated under this year's blanket of snow.

Later, I ask how fast Sperry is wasting away.

"How fast?" Fagre calls out. "How fast is it disappearing? We can only make a crude projection this summer." Perhaps earlier than 2030, as he has suggested before. But, after all this snow, I may have to wait for an update.

Dan catches up to Erich and Kevin and proposes they move onto the second phase in today's operation. They will test the "mass balance" of the glacier out of his bag of tricks, a technique that has vast advantages over measuring ice margins and calculating area. The latter misses one dimension—depth. A smaller glacier with more depth can be more massive, in terms of volume, than a shallow glacier with more surface area. You'd never know it just by looking at it. You just can't judge a glacier by its cover.

All other things being equal, the glacier with less mass (regardless of its area) will disappear first. Sperry might vanish before a smaller but thicker Grinnell.

The Europeans perfected the technique of mass balance to assess the gain or loss in a glacier's volume. The balance, either positive or negative or neutral, represents the *difference* between snow accumulation in winter and snow-and-ice melting over the summer. This arithmetic keeps track of the glacier's ice reserve, which can be likened to a virtual savings account, where accumulation represents deposits and melting represents withdrawals. An ice field with a sustained positive balance will yield interest—it will grow in volume. An ice field with a sustained negative balance will lose value—it will retreat. When the savings account is negative for many years, the ice becomes so depleted that the glacier cannot survive.

Fagre monitors mass balance only at Sperry Glacier, his favorite charge. With an eye toward sharing his results, he records the phenomenon in a universal measure. Mass balance is calculated in "meters of snow-water equivalent"—that is, snow and ice are measured in terms of their equivalent volumes of liquid water. This places all three on even footing. A bucket of snow may consolidate into a liter of ice or melt into a quart of water. Balances are measured every summer. You can discover a pattern by summing balance measurements over several years. The gain or loss is cumulative. Fagre and his researchers have been documenting mass balance at their benchmark glacier since 2005. Trends are starting to surface. And the patterns are not just earthbound. Mass balance has become a yardstick for measuring the health of the climate.

The three men walk in step across the snow, the rope stretched out between them, as if they were dangling on a necklace. The reds and greens and royal blues of their parkas stand out like tribal pendants. Erich taps the sides of his boots and crampons with his axe to release clumps of snow. When metal strikes metal, a *dring* rings out. They approach a white rod emerging from the snow.

The hollow white rod is an ablation stake, the main apparatus of a mass balance survey. Seven such stakes are strategically placed around the glacier: two near the top, three at midline, and two below—at the foot of the glacier. With up to seventeen hundred feet of elevation change from toe to headwall, the temperature difference may be 7.2 degrees F (4 degrees C) or more. This results in snow and ice being thicker at the head, thinner at the toe. Stakes are arranged to achieve an average. Placing them there, in June, was a prodigious feat.

At the summer solstice, what Fagre considers the end of the winter/spring accumulation season, the same team of three skied uphill to Sperry to document the snowfall for the year. Fagre led the way on his alpine touring skis. He orchestrates his team not like a conductor but like a first violinist who can play along with the best of them. Using long probes, they recorded a snow depth up to twenty-three feet (seven meters) or more. Probing has its rewards: The probes, made of aluminum but designed like bamboo, made a "thunking" noise when hitting the ice. This gave some reassurance that they were on target. The depth of the probes established the year's baseline. Next they planted the seven plastic ablation stakes to track the melting over the coming summer. Kevin hoisted a gas-fueled steam drill, which he had schlepped to the glacier, and perforated the snow until he struck ice. Then Erich and he threaded the holes with the ablation stakes. Each stake came in five sections, each one-and-a-half meters long—making seven-and-a-half meters total (over twenty-four feet). The tips peeked out of the snow. The sections were attached with black plastic zip ties, so that each section, standing vertically in the hole, would fall over, in succession, like a blind man's white collapsible cane, as the summer melting took place.

Now, in August, the melt level is measured. By subtracting the meters of melted snow from the highest snow level, the team will have the mass balance. This is like measuring how many gallons are left in a tank of gas: You subtract the fuel burned from the fuel purchased. The balance is what's left over or in deficit. Thus, Fagre would discover if there's a deficit or surplus of snow and ice—that is, if the glacier is out of equilibrium or not.

Erich taps his crampon again and it rings out with a *dring-dring*.

The glacier team reaches the first ablation stake, along the mid-section of ice. The white post is composed of one-inch-diameter PVC pipe, marked in one-centimeter increments along its length. Kevin kneels to measure the gradations on the stake. He marks the date and melt level on the post with a black Sharpie; Erich does the same in his field book. Two sections have fallen over, representing three meters of melt, approximately ten feet. Kevin warns to keep everything metric, the language of science. Snowfall was close to seven meters, so the mass balance in terms of snow is somewhere in the neighborhood of four meters—in the positive column. But this is only raw data. The team must convert the difference between accumulation and ablation to water equivalency. This will take some doing: They'll have to determine the density of the snow—up to two meters under the surface.

Dan Fagre offers to dig the snow pit. This is the least desirable job of the day—to excavate a truckload of snow from the glacier, creating a box-shaped hole. The purpose is to take samples of snow along a depth gradient from the surface to the bottom. It has to be a gradient to smooth out irregularities in the data. He's looking for an average of the density layers. The samples, each a liter in volume, are weighed on a miniature electronic scale. By dividing the weight by the volume, they could calculate the density of the snow. Water has a density of one hundred percent. Champagne snow—the fluffiest variety—has a density of six percent. Other snow densities—the moisture content of powder to slush—fall in between. By knowing the density of the particular snow, they could calculate the water equivalent of the snow column. Since the new snow sits on top of the ice, this tells you how much water is added to or subtracted from the glacier's mass. The density pit gets you there.

"The concept is simple," says Fagre, "but the acquisition and crunching of numbers are a chore."

We witness science in the making.

Fagre grabs a shovel, while Erich and Kevin head to the next stake. His first scoop rebounds with a thud. The snow is dense. Opening a red Swiss Army knife, Fagre strikes the snow, but it only sticks in an inch. With the shovel now, he flails at the snow, chipping off a few shards at a time. After twenty minutes, he's gone one foot down. Normally, he'd have dug the entire pit in that time.

"The densities are getting closer to ice than to snow," says Fagre. "The snowflakes have consolidated and refrozen so that the crystals are well on their way to becoming firn."

Firn is the previous season's snow that has thawed and re-frozen and compressed; it is more than halfway toward ice. On a glacier, the process of turning snow into ice takes two or three years and runs through several stages. Snow crystals change shape, consolidate, thaw, and refreeze. They succumb under temperature and pressure. On that journey, firn is the last incarnation before ice.

"In a normal year, we can measure a firn line as well as a snow line," Fagre continues. "The snow line is seasonal snow, from the immediate past winter, and the firn line is older, darker snow that's almost ice."

Both lines are measures of ice being built: The farther down the slope they are, the healthier the glacier is.

"Firn is money in the bank," says Fagre. "We don't know how long it'll last; it may transform into ice slowly or it may melt. But it's a great boost to the glacier right now. It's a relief to find firn in this pit. Sperry's on the rebound, at least for the year."

I'm taken aback by Fagre's optimism. Despite Sperry's dismal record of melting, he's still rooting for the glacier. He's the top cheerleader for the ice.

Having abandoned their boss to his digging, Kevin and Erich are halfway across the glacier. They have spotted the next ablation stake. Kevin is now the one to rap his crampons free of snow: *dring-dring-dring*. Erich waits for him to catch up. White sections are jumbled at his feet. After measuring, the pair stuff the discarded sections into their backpacks, which resemble baskets with overgrown chopsticks peeking out.

They move on. The documentation at each station is always the same. First task: Kevin marks the melt level and date with the black Sharpie.

"Measure for measure," one of them says.

"As you like it," the other quips back.

The two fall into an easy banter, punctuated by obvious and obscure literary references for amusement.

Measuring snowfall and snowmelt can be a chore. Both tallies must be entered into the GPS as well as the logbook before they move on. In the middle, a minor mistake is made, which must be blacked out with the Sharpie and reentered.

"A comedy of errors," somebody mumbles. The repertory is just getting warmed up.

The survey team moves quickly from stake to stake. Some are hard to find because the white pipe "disappears" against the backdrop of snow. But the pair has GPS coordinates to locate each rod. The snow depth varies drastically at each site. This uneven terrain demonstrates how untidy science can be: The glacier is not uniform but laced with piles of snow and shallow holes. These must be factored in. In June, Fagre intentionally placed a couple of stakes in an avalanche path to record the rogue input of snow. Sperry Glacier is not built by just snow flurries; wind pillowing and scouring and avalanche deposits are significant ingredients as well.

They can double the height like whipped cream atop a scoop of ice cream. However, he believes these ancillary snow sources do not have an "overwhelming effect" on the glacier's mass balance.

To verify the snow depth at each stake, the two men probe to the hidden ice surface. Sometimes a stake encounters an ice lens—and stops short—so the probes double-check the vertical profile to make certain it is pure snow. The team is meticulous in obtaining and reporting data. After all, their results will be scrutinized by critics of climate change. Their measurements and conclusions tend to be conservative. Their science must be above reproach.

Ablation stake "number seven" is located in the most dangerous place—right below the bergschrund (the frozen cataract that separates the stagnant headwall ice from flowing ice on the main glacier). It is besieged with icefall. But crevasses are still the most serious obstacle. Kevin belays Erich across the tricky approach, without a glitch. Then, Erich reels his comrade in. They record the melt—here a little less than four meters. They leave the stakes in place for another rendezvous in late September.

Kevin and Erich are pleased to be done. The errors have been corrected and the data are secure. Later, they tell me about their circuit.

"All's well that ends well," one of them says. They head down to the snow pit. I can imagine the repartee going far into the midsummer's night.

The average mass balance for the year will be added to the inventories of the past six years to create a multi-year profile of the glacier's gains and losses. This history will be summarized in the work of Joel Harper, University of Montana, and Blase Reardon, one of Fagre's graduate students, now living in Idaho.

Harper and Reardon have undertaken an eight-year study of

the mass balance at Sperry to illuminate how the dramatic ice reduction mirrors the new climate in Montana. To this end, they
have assembled mass balance and temperature records from 2005
forward. Preliminary results for the first two years are telling. For
the first year, the annual net balance was negative 1.22 meters.
For the second year, the net balance was negative 0.87 meters.
Low accumulation in one year (followed by a cool summer) was
trailed by high melting (following modest snowfall) in the next,
so the net balances were both negative. The average mass balance
for the two years came in just over negative one meter. Over
three feet of ice was lost from the glacier's thickness on average
each year. But remember: Mass balance is cumulative. For two
years, that's six feet, the height of a man. Imagine twenty years of
this wasting—that is, jump forward to 2033. If the pattern persists,
we will have lost sixty feet of ice mass. That is nearly half of Sperry's mean depth (147.5 feet; 45 meters). The arithmetic here is sim-
plified but the point is the same. The glacier will vanish. Or retire.
(A glacier less than one hundred feet thick is no longer active.)
Eventually, bedrock will be all that's left behind, except for a scat-
tering of ice, like cubes spilled on a kitchen floor.

The most intensive mass balance survey in North America has
been the North Cascade Glacier Climate Project in Washington
State, where the highest concentration of glaciers is found in the
lower forty-eight. (Glacier National Park comes in second.) Since
1984, ten glaciers—with names like Columbia, Ice Worm, and
Yawning Glacier—have been measured in the North Cascades.
Through 2009, the mass balance averaged negative 1.7 feet (0.51
meters), creating a cumulative loss of 43.5 feet (13.25 meters) in ice
thickness. For the ten glaciers, this represents a volume lost of
twenty to forty percent. The impact of the negative balances has

been that glaciers are not only thinning but moving more slowly. The lethargy of these giants may sound their death knell. A glacier that stops moving is just an idle chunk of ice.

Remarkably, the annual balance is very similar for all glaciers, which demonstrates that regional or global climate change is at work, not just an anomaly in local weather. The cause of the decline is primarily a rise in temperature. Ablation, or melting, has been a stronger impetus than curtailed snowfall. In fact, there has been a small increase in precipitation, and yet persistent snowpack has lessened. Cross-country skiers and others identified the reasons. The reduced accumulation is due to winter rain and unseasonal melting in winter.

It is worth noting that the annual mass balance loss at Sperry (negative one meter for 2005–2006) is double that of the Cascades, perhaps another indication of Montana's greater vulnerability during 2005 and 2006.

Two USGS benchmark glaciers in Alaska, the North Cascades group, and three Canadian ice fields make up fifteen North American glaciers that report to the World Glacier Monitoring Service (WGMS). Fagre tells me Sperry will soon make sixteen. The WGMS compiles and disseminates information on glacier fluctuations around the globe. Its database goes back to 1894 when glacier observations began with the intent of solving the puzzle of why ice ages formed. Since then, the goals of international data collection have evolved and multiplied. The WGMS is no longer just a window into the past. The future of glaciers is now on the docket.

More than thirty countries, ranging from Argentina to France to India, report mass balance data through WGMS, encompassing over eighty glaciers. The mean annual balance for their roster has been negative for twenty years (1991–2010). If this profile were a financial report, the company in question would be bankrupt. To streamline the balance sheet, thirty reference glaciers have been selected that mimic the range of values. For them, over the thirty years, 1980–2009, the cumulative loss in thickness per glacier is 41 feet (12.5 meters). To be sure, individual glaciers have waxed and waned: In Austria, the Hintereisferner receded from 1952 to 1964, then recovered, advancing until 1968. It then vacillated for ten years only to lose ice continually for the thirty-five years since. It has lost twenty-six meters in thickness since 1952. That is eighty-five feet, nearly the height of the cliff dive at Acapulco. Multiply this by the glacier's surface and you have melted the equivalent of a large lake.

One more statistic for good measure: If you group the mean balances of all the glaciers from thirty countries into decades, you find the values have declined from negative 0.66 feet (0.2 meters) in the 1980s to nearly negative 1.3 feet (0.4 meters) in the 1990s, to more than negative 2 feet (0.6 meters) in the 2000s. The decadal balance, the ten-year loss of ice, tripled in thirty years. This shows that climate disturbance is happening fast globally, and that it's pervasive and accelerating—like a building seismic wave.

Erich raps his ice axe against his boots once more, announcing his arrival to Fagre. Kevin stands close by. The sun, now unblemished by a single cloud, has softened the snow, causing it to ball up more on Erich's crampons. The ice axe does the trick: *dring-dring*.

Fagre climbs out of the pit, using the snow stairs he has left in place. Excavating the hole was like digging out a stairwell, revealing a series of snow steps to the bottom. It had taken two hours to uncover the well. Once on top of the surface of the glacier, Fagre scans the horizon and lifts his eyes in the direction of the last ablation stake just above them. He turns to his crew and raises one brow. The field report comes fast.

"We got all the measurements," says Kevin quickly. "Everything was clean, Boss."

"Good. I'm nearly done with this snow pit."

Inside the hole, Kevin and Erich employ a special cutter to extract one-liter cylinders of snow from the uphill wall. More than a dozen samples are taken. Afterward, that series of holes resembles an empty wine rack. Each sample is weighed on the electronic scale and the heft of it recorded. Weight will be divided by volume. The measures will be averaged into a single density to convert the accumulation and ablation values into meters of snow-water equivalent.

Just above the hole, Fagre munches on a peanut butter and raspberry jam sandwich while his men wrap up the measurements. "A working lunch," he says. Fagre won't take an hour for eating a meal or painting a canvas or any other hobby besides staying fit. He's in a race against time to document the glaciers before they vanish. That race is fueled by his motivation to be professional and near perfect in gathering his profiles of the ice. By tracking the melt accurately, he offers a trusted warning to the world. But underneath this drive, and at odds with it, is the wish that perhaps the glaciers will somehow survive.

Back at the lab, calculations will reveal that Sperry's mass balance for the summer is positive in snow-water equivalent. The positive net balance confirms the glacier has improved for the season. Fagre is jubilant: Sperry has a chance, albeit a slim chance, to pull through.

"The snow line—really the near-complete snow cover—gave it away," Fagre says, "but it's good to have the positive value confirmed with a precise number. Now we can track the balance to see how it meshes with subsequent and previous years. Sperry may give up the gains or it may be the spark of a new trend. We'll see. Climate and glaciers measure time in decades."

The big news about glaciers usually depends on extrapolation: What is good for the goose is good for the gander. By this reasoning, the six other glaciers under Fagre's gaze are likely positive for August, too. Typically, they follow Sperry's lead, for the premier glacier probably has the strongest climate signal in the Park. It is the bellwether, the most responsive to changes in temperature and precipitation. Still, he would like to take their measure directly. But Fagre cannot harvest any more data with all this snow.

"Like a farmer who didn't get his crops in—that's what I feel like," he says later. "Now I may have to wait for another season."

Unscathed for the day, the three mountaineers return to the rock rib to regroup. Once on hard ground, they unburden themselves, releasing their gear. Fagre unbuckles his harness and drops it like a gun belt to the ground. Erich untethers his crampons, placing the blades upright like a lethal metal rake, then thinks better of it and ties them to his pack. Kevin coils the climbing rope, which is as cooperative as a garden hose, full of kinks and figure eights. The gear must be stowed with care, compressed and compact to thread the narrow needle of The Staircase along with the team.

The men, now in bare boots, have one last task before departing the glacier: checking the weather station. Kevin confirms the battery is working and being recharged by the solar panel—always a relief for a remote outpost like this. He checks the five sensors, one for each kind of measurement—temperature, wind speed and direction, solar radiation, among them—and finds the wind gauge dead. He quickly replaces the sensor. All five measures are now running like a pentathlon.

"The wind can top a hundred miles per hour, blowing across this ice in winter," Kevin says. "Even in summer we've been caught in storms up here that were blowing plenty hard. Lightning, too. On days like that, I keep telling myself I chose this job for a reason."

"Sperry Glacier is not here by coincidence," says Erich. "It's here because this cirque gets bad weather. Glaciers thrive on heavy weather. I guess we do, too."

"You gotta love the mountains, love glaciers, to trek up here," says Kevin. "That's for sure. Heavy weather or not. A sunny day like today makes this job a walk in the park."

Fagre's other five weather stations, placed strategically in the alpine zone around the Park, from Logan Pass to the Garden Wall, will function all winter. Their output will be examined to find, among other phenomena, if more rapid warming has manifested at higher altitudes. Lindsey Bengtson is currently collating the weather feedback from all these stations. She's risen through the ranks. To land her research post, to prove she could do it, she, like so many others, hauled countless car batteries up to high-altitude weather stations. After seven years of toil, Fagre was impressed by Lindsey. "Now, I let rookies haul them instead," she says. "It's amazing how many martyrs there are."

Kevin, who is standing in for Lindsey today, stoops over to tighten the loose guy wire that was discovered earlier this morning. It only takes a couple of minutes. Kevin and Erich look at each other and nod their heads. The repair list is complete. And the glacier fully examined.

Kevin unlaces his leather boots and removes them. He replaces them with running shoes and stashes the heavy boots with his crampons in a haul bag underneath some rocks.

"I'm not much of a hiker," he says. "I'd rather run or ski down to the car. Well, I also need a break from those confining boots. My feet do stink."

That's Kevin's final experiment of the day: To see if the marmots are attracted to or repelled by his stash. My bet is they chew his boots down to a nub.

Fagre agrees all chores are done. "We've worked up the vitals," he tells me later. "We must leave the rest for another day. Melting will continue for a few more weeks, but October will bring more snow fast. Then it's a long winter. Basically, in Glacier, we have ten months to accumulate snow and two months to melt. That's why I love winter. It delivers snow to the glaciers."

I take a deep breath before contemplating the descent and scan the ice field one more time. It seems vibrant but that is an illusion—a sleight of hand due to all that snow. On a year like this, the sun's rays are still melting a fair share of ice and snow at the foot of the glacier at summer's end. A thousand tiny capillaries of water flow along the surface and combine into arteries under the heat; as a result, the glacier withers. (The coming Indian Summer, in fact, will bring a warm September and push the melting season into the autumn: All that snow will be no match for the heat; it will turn out to be a negative balance for the year after all.) The outflow streams now rush with meltwater toward the precipice, that three-thousand-foot cliff above which Sperry is perched.

Fagre tells me that the streams are home to organisms that are only found in the Park's glacial creeks and nowhere else on Earth. I promise myself to take a look on another day, but at the moment I head down the mountain far ahead of Fagre and his crew.

At the top of The Staircase, I change footwear—out of boots, into trekking shoes—to be more sure-footed. The descent is twice as unnerving as the ascent, facing outward with my back to the wall. The steps are elusive, out of sight. I must feel my way with my toes.

After that, the rest of the ten-mile hike out is a cakewalk. Alpinists are known to say that the most beautiful views are atop the toughest routes. I think that's true. I've seen some vistas in the Alps and Alaska that are as stunning as the climbs are challenging. The panorama is staggering: Mountains that simply go on forever, as if wilderness were my entire world. Glacier National Park has climbs like that. But I'm happier on the long, grinding trail today than on the stairway to heaven. I'd rather have a blister than vertigo. One is annoying, the other unsettling if not terrifying at times.

Such inconveniences vanish overnight. I'm left with the memory of the day. Visiting the glacier is like any relationship, I have discovered: It requires passion, a little sacrifice, and the capacity for joy. In return, you get kinship and a partner. Today, I feel less alone, and even more aligned with the community of glaciers. I'm invested in the ice.

The Cascade Effect

On the way down from Sperry, I pick up a small rock, cut to the size of a gold nugget and polished smooth by the enterprise of time. I drop it in my pocket for good luck. In Glacier, it never hurts to call on the angels. While not famous for its bears like the Grinnell footpath the trail through the downward-sloping forest below Sperry Chalet looks like prime bear habitat to me. And I can't look around every tree. I'm in a hurry. I want to arrive at the base of the exit falls, eight miles below, by mid-afternoon. I'm taking the long way down—navigating a hundred switchbacks and loops. The trail curls around the mountain like a corkscrew.

My knees feel like rubber bands.

Without incident, I reach my objective by two o'clock. I spy the long cascade of Avalanche Falls through the hemlocks. Next minute, I'm standing near the lake of the same name: Avalanche Lake. The cool pool—perhaps a little warmer these days—receives the meltwater from Sperry Glacier via the single, spectacular waterfall, a three-thousand-foot descent. Because Sperry is smaller than it was, the waterfall is usually a trickle—a light if long out-door shower. But the third Monday in August, the day after our climb to Sperry, the sun is bright and the water gushing. Fagre tells me the glacier has become an enormous reflector oven, con-centrating the heat within the cirque, with its crystals of ice and snow. The temperature has climbed—pushing 80 degrees F (26.6 degrees C). Imagine everywhere across the glacier small rivulets of ice water manifesting—like sweat—then flowing and zigzag-ging, joining others while gathering speed. They rush toward the foot of the giant ice field, approaching the toe, the precipice. They align into a single stream. Near the edge, the waters compress into a funnel. The chute draws all moisture from the glacier run, just as Niagara does from the St. Lawrence. The big drop is the only way out.

The ice water falls so fast and so far it actually warms up on descent.

I have brought a map and compass (and a thermometer) along with my notebook and an assortment of pens. I am searching for the downstream impacts of Sperry's decline. Fagre warns me that the ebbing glacier and warming cascade are only half the problem. The glacier is melting—this is certain—but the downstream con-sequences, acting like a series of domino-like, cascading effects, are ubiquitous. And yet poorly understood.

Those impacts are intertwined like the intricate web of a spi-der. Touch one strand and the hub can come unglued. Climate is integral to the functioning of ecosystems, but any change in cli-mate can bring about stress that is disruptive to the community. Take early snowmelt: Winter flooding erodes stream banks, dis-placing plants, wildlife, and fish. The timing of flowering and spawning is also derailed. Snowfields that traditionally have stayed frozen until spring or summer have been on the decline in Montana since 1951. On average, over these past sixty years, winter snowpack has dropped by seventeen percent, in depth and extent. That's nearly two feet lost out of ten. Imagine a house without its roof—that's the magnitude of the loss. The snowfields are much more extensive than glacial ice and still contribute the majority of spring streamflow throughout the state. River levels have trended lower of late. Snow and ice are rarer commodities.

Mountain snowpack has been called an alpine "water tower," storing moisture high in the watershed and then slowly releasing it when water is needed the most. Meltwater traditionally flows in modest pulses through spring and summer, even into the fall. An alpine gift but also a form of damage control. Blanketing the slopes sometimes by ten feet or more, snowpack is a protective reservoir, without which—that is, if winter precipitation only fell as rain—the alpine slopes and valleys would be flooded all winter long, with untold injury to the landscape.

Water is humanity's most basic commodity, its storage and release paramount to each village, each town, each corner of civilization. Globally, glaciers and snowpack combined are responsible for the majority of freshwater supplies. In many locales, distant alpine glaciers and adjacent snowpack fill lowland rivers with the base flow upon which agriculture depends. The near disappearance of the Zongo Glacier in the Bolivian Andes has already created an irrigation crisis for downstream planters. I write in my notebook: Could the farmers of the Missouri River or those of the Saskatchewan, or the Columbia, be next? From Triple Divide Peak on down, these three watersheds are fed by less snow each decade.

Throughout the American West, river flow is dependent on winter snowpack. Some estimates of its contribution exceed seventy-five percent in the states adjoining the Rockies, Sierras, and Cascades. All these mountain ranges are suffering a sharp decline in snowfield coverage. During the winter of 2009–2010, the mountain snowpack for the Flathead River, which drains Sperry and Avalanche Falls and feeds the Columbia River, was seventy-eight percent of average years. This translates directly to streamflow. The peak flow of the Middle Fork of the Flathead River (near West Glacier) was fifty to eighty percent of the average from 1971 to 2000. We have a trickle-down ecology. Tourism is down—it sinks with the water level like a cork. Whitewater rafting companies, among others, complain. They have fewer clients. To whom can outfitters and farmers turn for remedy?

A raindrop lands on my open notebook, splattering the ink. The tip of my pen swims in the aqueous solution, now useless for writing. This is the culprit, I think: the raindrop. It plays beast to the beauty of the ice crystal. This is the balance between summer warmth and the frozen winter. The balance has defined our climate until recently and nurtured each biome and ecosystem in the temperate world. Now that rain has surpassed snow as the major component of annual precipitation, the water towers are lean. The reasons behind this imbalance could theoretically fall to one or more climate factors: local temperature anomalies, natural variability in the climate (Pacific Decadal Oscillation, El Niño, La Niña, etc.), or human-induced global warming. Amidst a heated debate, the majority of scientists believe global warming is in play, while a minority holds that only climate variability is responsible. Yet the immediate role of temperature in bringing down the rain is undisputed. The temperature is warming; the debate is only about the cause.

From 1951 to 2006, Montana's average air temperature climbed

2.3 degrees F (1.3 degrees C), compared to a 1.25 degrees F (0.7 degrees C) increase for the entire United States. Spring rain has increased 5.9 percent, leading to an earlier snowmelt. The two measures over five decades—temperature and precipitation—comprise climate change. Montana, which has much territory at elevation along the Canadian Border, has shown 1.8 times as much temperature gain as the rest of the world since 1900. Among several reasons, higher altitudes are often blanketed with heat-trapping clouds, and the jet stream brings in warmer winds from the west.

Scientists also agree that elevated temperatures are the major force behind a lighter snowpack. An occasional rise in snowfall, for a winter like 2010–2011, has been insufficient to overcome losses over a given decade. In any case, rain now rules. In Glacier recently, it has rained in January—three years in a row. Moreover, any remaining snowbanks have been overwhelmed by early melting. Across the West, the peak snowmelt date has advanced ten to thirty days. Early flooding is the new normal. Streams are low in late spring; rivers even lower in summer.

The warm phase of the Pacific Decadal Oscillation, such as was found during 1977–1998, which brings on warm and dry winters in the Pacific Northwest and Montana, can explain about a third of the warming at Glacier, says Fagre. But only a small fraction of the additional precipitation can be attributed to any of the Pacific climate indices. Meanwhile, temperature and precipitation increases are very consistent with the global pattern of man-made warming.

Dan Fagre reminds me that drinking water and irrigation are not the only beneficiaries of ice and snow. On the Flathead Reservation to the west of the Park, most of the tribe's energy is obtained from hydropower. The turbines turn at the edge of a reservoir. The reduction of meltwater threatens their potential to generate electricity.

But these are only services to humanity. The natural benefits of ice field and snowpack to alpine ecosystems, species, and habitats are fundamental and far-reaching. Snowbanks furnish winter dens for wolverines, grizzly bears, and other hibernators. The lynx and its preferred prey, the snowshoe hare, romp across the surface of the snow. The hare depends on snow for camouflage. Winter birds like ptarmigan also employ cryptic coloration against the background of white. In all, a wide array of the furred, feathered, and finned relies on the snowpack for sustenance. They are creatures of winter.

My eyes wander overhead, from the high, hanging bench of Sperry Glacier, down along the verdant waterfall, to a little stream at its base that meanders past willows to the lake at my feet: Avalanche Lake, home of one of the most imperiled aquatic species in Montana. The teardrop basin is 1,300 feet (396 meters) across. Driftwood and submerged logs ring the shallows like barbed wire around a pen. Beyond the debris, at mid-lake, the water surface is smooth and unburdened, a broad mirror reflecting the image of the cliffs. The steep basin is a cul-de-sac: three rock sides on the scale of Yosemite and a cobbled exit for Avalanche Creek, which drains the lake. No fewer than five waterfalls crash from ledge to ledge toward the lake. Sperry feeds just one, which only lasts till summer's end, while the Continental Divide, high above to the east, fuels Monument Falls, which keeps churning into the autumn. In long scars adjacent to the falls, a dozen avalanche tracks are carpeted in green—grasses and shrubs and subalpine fir that may cling there until the next rush of ice and snow comes by. An ice freight train on schedule for descent each spring. The broken trunks and rubble below each chute speak to the annual violence.

Besides this cacophony, the lake is tranquil for most of the year. It is cradled like a sparkling gem between the rock walls. In winter, it is snowbound in solitude; in summer it is a watering oasis—for wildlife, for people. All the time, underwater life survives. Just now, a male common loon—a diving bird and fish-eater—makes a series of falsetto calls from the east side of the lake, best described as a manic yodeling. His laughter is my first clue that there are fish in the lake.

Closer to me—say, fifty yards away—a single splash, really a pinprick, punctures the glass surface of the lake. A tiny ring of ripples pulsates outward for a few feet, a concentric wave, then gives up. At first, I think the disturbance might be another raindrop, but when I scan overhead, the sky has cleared into the cobalt of August. Suddenly, three more ripples—if not a school, then at least a gathering of trout.

Despite lower water levels and a warmer basin, the westslope cutthroat trout *(Oncorhynchus clarkii lewisi)*—the Montana state fish—still survives in Avalanche Lake. Like the threatened bull trout *(Salvelinus confluentus)*, which also occupies frigid lakes and streams just below glaciers in Montana, the cutthroat is a coldwater species, intolerant of warmer temperatures. With Sperry and its snowfields discharging less cold water, Avalanche Lake is warming up. The trout are in trouble.

Westslope cutthroat trout occupy a few other watercourses within the Park, but many populations are said to be "fragmented." Fish are segregated in patches of habitat, essentially aquatic "islands" in the streams. For example, cutthroats may reside in two or more forks of a tributary but not in the main stem, forever isolated from each other. The trout swim higher and higher—from river to creek to stream to headwater—looking for cooler waters. The runs become narrower, shallower, obstructed, and eventually become dead ends. Other forms of habitat degradation, such as erosion and siltation after wildfires, further harm the watersheds. For all trout species, nearly thirty percent of their habitat has been lost in Montana.

To compound problems for the cutthroat, an exotic species presents a second risk to the fish. The coastal rainbow trout *(Oncorhyncus mykiss),* not native to Montana, has been introduced to state waters—for anglers—to the detriment of local species. Rainbows pose an insidious threat: They have diluted the native gene pool. The rainbows crossbreed with cutthroats, producing hybrids that are somewhere between the two.

Of the many consequences of crossbreeding, one of the most self-destructive for the original stock is behavioral. Pure cutthroat trout are natal spawners, returning to their birth streams to breed. However, hybrids lose much of the homing instinct, so they spread out, further corrupting the genome elsewhere in the watershed. Like blood in a pool, the genome dilutes all the more the farther it is from its source.

Hybridization, the mixture of genes from two species, is a result of interbreeding. Most animal species have isolation mechanisms— for example, different habitats or breeding behaviors—that prevent hybridization. When it does occur (for example, between a donkey and a horse), the offspring (in this case, a mule) are often infertile. Crossbreeding fish are different: The progeny are sometimes viable. In fact, hybridization is more frequent in fishes than in any other vertebrate group. Trout, like many other fish species, have external fertilization—namely, spawning (releasing eggs and milt) into riffles in a stream. This readily permits interbreeding. Introducing exotic species can eliminate barriers to hybridization. The shrinking of pristine, coldwater habitat encourages non-native fish like the rainbow trout to invade and interbreed.

Rainbows, which are native to the Pacific slope (Western Mexico, California, Oregon, Washington, and north to Alaska), have become the most widely introduced trout in the world. (As a teenager, I caught exotic rainbows with my hands out of the Wind River in Wyoming, a practice called "skulking.") For over eighty years, fishery managers intentionally stocked rainbows in cutthroat habitat throughout the West. They did not imagine the downside to the practice; they were only concerned with sport. The cutthroat was less popular among anglers because it was too easy to catch. Not only was it easy prey for the fisherman, it proved to be a poor competitor to exotic species. It couldn't stand up to the rainbow. Even as cutthroat stocks plummeted, management continued to favor rainbows, thanks to their leaping ability on the hook. Consequently, rainbows have been stocked in all three drainages of Glacier National Park, where cutthroats once reigned. Pure cutthroat now prevail in perhaps less than ten percent of their historic range.

Introduction of exotic species now ranks third only to habitat destruction and pollution as the basis for extinction in modern times. Sometimes native species are displaced; other times, they are hybridized with the invader and lose their genetic integrity. For example, mallard ducks have "swamped" the gene pool of Florida mottled ducks, reducing the identity of the local species. Recently, there have been various reports of extinction from hybridization, particularly among small populations of native plants on remote islands like Hawaii and Guam. Hybridization induced by human manipulation or error is fast becoming a major cause of species loss worldwide.

Native trout could be next.

After nearly a century of stocking exotics, most of it random and unrecorded, the genetics of cutthroat trout are permanently muddled. The mixing of genes is irreversible. While rainbows and cutthroats produce fertile offspring, the outbreeding often continues, generation after generation, until a hybrid "swarm" is created, and the native cutthroat genes are lost. At some point, the cutthroat is no longer a cutthroat. In the future, it may be declared extinct.

Cutthroats demonstrate a low tolerance for temperature increases; their lethal water temperature is 67.3 degrees F (19.6 degrees C). Meanwhile, rainbows have a very high tolerance for warmth. Their lethal temperature is 75.7 degrees F (24.3 degrees C), approximately 8.4 degrees F (4.7 degrees C) higher than cutthroats on the upper end. This dichotomy makes hundreds of riverine miles more accessible to exotic fishes. Consequently, rainbow trout have a distinct survival advantage over cutthroat trout in warmer rivers and creeks. Hybridization is thus more common in warmer and degraded streams. There they interbreed with any vestige cutthroat population, corrupting the gene pool. For this reason, global warming is promoting the genomic extinction of the westslope cutthroat trout in the majority of its watersheds. Only in isolated, cool pockets with natural barriers to rainbows, like Avalanche Lake, may pure cutthroats survive.

But why should we preserve the genome of the cutthroat trout? Why should we care? Why should we rein in climate change for a fish? Preserving biodiversity, the variety of life, includes saving each ecosystem, habitat, species, and gene pool. To paraphrase Aldo Leopold, the father of modern conservation, every cog and wheel plays an important role in the workings of the whole machine. No creature exists in isolation, and removing just one can have unforeseen, calamitous effects. Some are flywheel species, absolutely key to the machinery; others are minor gears. Honoring the planet, that wondrous machine, begins with local works— each backwater pool and stream, each trout and loon, each glacier on a mountain horizon. Many other reasons are given for looking after biodiversity, namely all the benefits provided to humanity: agriculture, food, medicine, pollination, air and water purification, soil fertility, housing and clothing, carbon storage (e.g., trees and permafrost), flood control, and more. We all depend on the services and products of ecosystems. But, for many of us, each species has an intrinsic value—separate from any service to people—and, thus, to exterminate a species that has developed over millions of years is not only shortsighted but arrogant and wrong. Nonetheless, forty times more American species have gone extinct in the past century than in an average century in the Pleistocene. We are to blame.

"Before, we took a Band-Aid approach to saving species," says Clint Muhlfeld, forty, a USGS aquatic ecologist. "Protection and recovery were focused on small-scale efforts. Climate change has altered our outlook and our resolve. Now we are looking at entire ecosystems, the full watershed and beyond, and these over longer periods of time. Global warming knows no boundaries, so scientists and managers must widen their scope, too. That's the only hope for the cutthroat and other vulnerable species."

In 1997, a coterie of environmental groups petitioned the U.S.

Fish and Wildlife Service (USFWS) to list the westslope cutthroat trout as "threatened" under the Endangered Species Act (ESA). American Wildlands, Trout Unlimited, and a number of other organizations warned that the cutthroat was likely to become "endangered" within the foreseeable future. Endangered status is designated when a species is in danger of extinction; threatened is one rung down. The petitioners claimed the cutthroat was threatened because of habitat degradation and hybridization, competition, and predation from non-native fish. Global warming was not identified as a threat; these were early days for climate change awareness. Nonetheless, environmentalists and bureaucrats bantered back and forth on the habitat issue. The main argument for listing the species in the government's eyes was the loss of more than forty percent of its historic range, their low-end estimate. The exact percentage was under contention. Some experts claimed the figure was closer to ninety percent. Fish were largely restricted to fragmented headwater habitats that were vulnerable to wildfires and flooding; more losses were imminent. After the first volley, the USFWS shelved the petition. To force their hand further, the co-petitioners initiated two successful lawsuits to prompt the government to act on the proposed listing. Three years later, with fire at their heels, they officially considered the petition. And rejected it. The government claimed the presence of westslope cutthroat in 23,000 miles of streams and rivers in the U.S. and Canada did "not warrant" listing under the ESA. In other words, they asserted the species had an adequate contemporary range.

Ralph Morgenweck, then a regional director of the USFWS, said, "Viable, self-sustaining westslope cutthroat stocks remain widely distributed." He pointed out that they swim freely in Montana, Idaho, and portions of Oregon, Washington, British Columbia, and Alberta.

There's a huge catch here—essentially a "Catch-22." The USFWS included hybrids in their count of cutthroat populations. By adding "cutthroats" with up to twenty percent rainbow genes to the tally, they overstated the range of the pure species. The irony is that the initial petition flagged hybridization as one of the major threats to the species, and then, come judgment day, hybrids were employed and counted as the reason the species was safe.

Rob Ament, executive director of the lead petitioner, American Wildlands, said at the time of the rejection, "You can't count hybridized populations to say the fish is stable. The numbers are inflated." The arguments and counterarguments for and against listing the fish form a patchwork of responsibility and denial.

Silver threads in this fabric are the efforts of Montana's Fish, Wildlife, and Parks Department to protect cutthroats by restoring stream habitat. Recovery of listed species is the goal of the ESA, so the State of Montana is pinch-hitting for the feds. But how do you screen out hybridization or global warming? There is no single-source polluter to shut down, no spigot to cut. These forces move into and through the riverine habitat as freely as the water itself. One measure put in place to curb crossbreeding is the deployment of steel barriers along select streams. Another strategy is the translocation of pure cutthroats to isolated streams and lakes. Hybrids are also being eliminated, with moderate success. But global warming is ubiquitous and long-lived: How do you shut off the heat? It pervades every corner of Montana, every reach of the planet, and, once past the tipping point, it may be impossible to hold at bay. The fishermen of Montana may be no more successful than the Dutch will be in holding off the next meter of the sea.

Aquatic species in Montana have been protected under the ESA before, most notably the bull trout, the top native predator in Rocky Mountain streams and lakes, and in the Pacific Northwest. Historically, Glacier National Park hosted one-third of the natural, undammed range of this species—until the invasion of the lake trout *(Salvelinus namaycush)*, a predator and competitor of adult and juvenile bulls. Habitat degradation has further prompted a decline. Local warming may affect the species, which spawns right below glaciers and perennial snowpack. However, climate change was not considered in the ESA petition. In 1998, the bull trout was still listed as "threatened" under the ESA, largely because it was considered secure in just two percent of Montana's clear, cold stream segments. Further studies will perhaps reveal why so few persist.

One domino falling from the decline in bull and cutthroat trout, which normally patrol the shallows, is the effect on foraging bald eagles and ospreys. The birds of prey have a chance to hook native fish along the lakeshore, but the deep-residing lake trout are largely out of reach. The food chain loses a link.

Breaking with an informal policy of keeping climate change out of the listing process, the USFWS considered a petition in 2010 to protect a local stonefly *(Lednia tumana)*, native and endemic to the Park's coldwater streams, under the ESA. The tiny aquatic insect, less than half an inch long, lives in frigid creeks fed by glaciers or snowpack (and in high alpine springs). The nymphs cling to rocks underwater and eat algae. The adults have wings and swarm above the streams, possibly becoming prey to black swifts that nest behind waterfalls. (Bears may also prey on the swarms.) If the ice and snow disappear, so goes the stonefly and its associated community and habitat.

Informally called the meltwater stonefly, the species prefers streams right below glaciers, where the streamflow is coldest. They have little tolerance for heat. Alpine streams warm up quickly away from their sources, so a couple of hundred yards downstream the stonefly may be absent. (Glacier-fed creeks are typically colder than snowpack streams so may hold a population longer in their grip.) The maximum summer afternoon temperatures of these frigid creeks seldom exceed 59 degrees F (15 degrees C). But now they are likely to get warmer, even at the headwaters. Or dry up completely. The available habitat is expected to contract by over eighty percent under climate change scenarios.

On a recent field trip to measure the lethal temperature for the stonefly, the threshold level at which its habitat would become unlivable, I saw many nymphs in one of their healthiest domains— Lunch Creek near Logan Pass, the central hub of the Park. The temperature experiment was organized by Clint Muhlfeld and Joe Giersch, forty-two, a USGS entomologist, both colleagues of Dan Fagre. We climbed to a spot on the creek, right below the snowpack, and were greeted immediately by a swarm of stoneflies. The nymphs had hatched into adults, their annual metamorphosis. Clint and Joe collected nearly two hundred nymphs off rocks in the stream. These were taken in iced-down coolers to waiting trucks for the flat-out drive to the lab. Hopes for the experiment were nearly dashed by a construction roadblock, but Clint calmly rolled down his driver's window and announced, "We have endangered bugs in here—let us through!" We were quickly waved on. At the lab, the insects were treated to an incremental temperature bath. The chronic lethal maximum temperature was 68.4 degrees F (20.2 degrees C) for stoneflies acclimatized to 59 degrees F (15 degrees C), a common creek temperature. At that level, they would die over time. It was important for scientists to know this threshold. Suitable habitat would only be found where it is considerably cooler than that.

Ecologists call the realized livable zone the animal's niche, or "climate space." Without water supplied by glaciers or snowpack, the climate space foreshortens or simply disappears. "The stoneflies can move upstream toward the snow," says Muhlfeld, "but they quickly run out of water. They hit ice or just dry rock. They have nowhere to go."

This dead end is called an "elevational squeeze." The squeeze is nowhere more pinched than at Sperry Glacier, right above me. Take the exit streams below the glacier. Meltwater stoneflies reside there, though not in the glacier ponds—they're too warm. So, here this ephemeral insect is caught between the ice and the plummet to the waterfalls—a distance of 250 yards, or so—just a patch of cool water. They would find the cascades too violent and too warm.

The stonefly petition also moved the USFWS away from what critics have called a preoccupation with "charismatic megafauna," that is, snowy owls and pandas and other warm and fuzzy creatures. These big animals have been valuable rallying points for action and education but tend to overshadow smaller, less cuddly species. And yet, the little creatures may be just as important— linchpins in the environment. Uncharismatic "microfauna" like stoneflies and trout often form the foundation of an ecosystem. Pluck one out of its habitat, and a whole community can collapse. As "keystone" species or simply members of the community, they are essential players in the ecosystem.

To its credit, the USFWS did not shrink in considering global warming as a threat to the stonefly's habitat. Nevertheless, in

April 2011, it declined the nomination. In the end, the government considered the petition "warranted but precluded" from approval because higher priority candidates were in the pipeline. Muhlfeld believes there wasn't yet enough information to make the case for priority listing, but that the ecological profile of the species is now in hand, so re-petitioning is on the near horizon. In the second round, he says, the stonefly's future will be clearly linked to the fate of the glaciers. The loss of one could easily mean the extinction of the other. A likely spinoff of this acknowledged connection will be the identification of the stonefly as an "indicator species" for climate change. If and when the population is further reduced, the impact will likely be indicative of climate-ecosystem damage. Similarly, because they are so dependent on high-quality habitat, bull trout and cutthroat trout are also indicators of stream health. Introduced species such as brown, lake, and rainbow trout are able to survive a wider range of environmental conditions; they are not as sensitive. In the wake of global warming, the natives will be the first to go.

Bringing them back from the edge will be the biggest challenge. If the meltwater stonefly is placed under protection of the ESA, the USFWS will be charged with restoring its habitat. This sets a high bar. Muhlfeld has doubts it can be done for the stonefly or any other creature vulnerable to climate change. "What can we do for the polar bear now that arctic ice is melting?" he says. "How do we bring back the sea ice? The same question is posed by the stonefly: How do we bring back alpine glaciers? It will require restoration—namely, curbing carbon emissions—on a global scale."

Administrative officials have also voiced concern over the restoration dilemma. "It's going to be very difficult to recover a species threatened by climate change," says Ann Carlson of the

USFWS. "We may have an invertebrate facing loss due to elevated temperatures. It may warrant listing. But that doesn't mean we can regulate coal mining."

Could an aquatic insect in Montana stop the building of a power plant in Seattle? Or New York? In the 1970s, the snail darter, a small endangered fish, nearly halted the construction of the Tellico Dam when it was discovered on the Little Tennessee River. The river and the darter were about to be inundated by the dam's reservoir. Despite an injunction against completing the dam, which was affirmed by the U.S. Supreme Court, Congress eventually exempted the dam from the ESA. The snail darter and the river lost. Distances are greater from the stonefly to sources of electricity, and the cause-effect relationship is not as immediate as it was for the snail darter. Still, it's interesting to consider whether the stonefly, *Lednia tumana*, could be the next "snail darter" in the public eye.

Pointing to the climate as the culprit in the stonefly's decline has only been possible recently. "A few years ago, we had very limited scientific information on global warming at the regional scale," says Ann Carlson. "But over the past year, lots of local temperature analysis has been going full-steam ahead." Much of the local data that indicts global warming in the case of the meltwater stonefly was collected by the U.S. Geological Survey. From temperature and precipitation data to snowpack and streamflow measurements, the USGS has assembled historic and contemporary information that profiles glacier-fed streams. Clint Muhlfeld leads the local effort to analyze the ecological requirements of the meltwater stonefly, bull trout, and cutthroat trout.

Muhlfeld occupies the building next to Fagre on cabin row. A small sign announces his division: "USGS Aquatics." Clint's mis-

sion is to assess the health of freshwater habitats and species below the glaciers and snowfields. He is thin and strong—built like
a swimmer, I think, but it turns out he's a bike racer. Like Fagre,
he has close-cropped brown hair. Like Kevin Jacks and me, he
spent his childhood summers on a lake in the Adirondacks. Exploring the woods and fishing the waters led him to a career as a
biologist. He splits his time between hard science and environmental advocacy. One of his current research priorities is the
question of how hybridization among trout affects the ability to
survive.

In the last decade, Muhlfeld and his colleagues have made a
remarkable discovery: The mixed offspring of cutthroat-rainbow
hybrids tend to produce, upon breeding themselves, fewer progeny. Later-generation hybrids are less fecund. This poor reproductive success has major fallout for cutthroats ecologically. In the
Darwinian sense of the "survival of the fittest," the hybrid offspring are less "fit," less likely to reproduce as a species. In the short
term, they are less likely to compete successfully with the other
species. In a long-term, evolutionary sense, the cutthroat hybrids
are more likely to go extinct.

"This is a surprise," says Clint. "Sometimes, hybrid populations are more fit than their parents, thanks to a phenomenon
known as hybrid vigor. But we're not finding hybrid vigor in fish
populations. Not in Montana."

Muhlfeld has surveyed the spawning success of a sample population of cutthroat intermixed with rainbows in Langford Creek,
a tributary to the North Fork of the Flathead River. He can spot a
hybrid in a net from coloration and other signs. With the aid of
genetic markers, he genotyped 61 females, 124 males, and 648
emigrating juveniles. By comparing their genetic fingerprints, he

identified the parentage of each juvenile, exactly which adult spawned each little fish. Those parents were then classified by the number of living offspring and by the proportion of rainbow genes. Reproductive success was calculated as the quantity of off-spring per parent for each spawning year. The results are stunning. The proportion of rainbow genes—even a small percentage—has a powerful negative effect on fitness. Reproductive success declines by approximately fifty percent, with only twenty percent rainbow mixing. This level of "introgression" (gene flow between hybridized populations)—i.e., twenty percent—was permitted in the census of cutthroat populations by the USFWS. Therefore, a fair portion of the government's "cutthroat population" must have contained hybrids with depressed fitness. Some scientists and environmentalists believe these weakened hybrids aren't worthy of listing under the ESA, but Muhlfeld believes any vestige of *pure* populations do warrant protection and recovery of the original species. He speaks with urgency. The waters are warming and the timepiece is ticking. Every year, more cutthroat are corrupted by rainbow genes and are thus less likely to reproduce. Moreover, the integrity of the species is in question.

"We found a paradox, however," Clint Muhlfeld is quick to qualify. "Introgression reduces fitness but proceeds rapidly through the population." One would expect hybrids, with their low reproductive capacity, to die off. But they don't. A partial answer to this conundrum may be found in the discovery that first-generation hybrids, as well as a few atypical males, have relatively high fitness, compared to later generations. This may afford frequent backcrossing, the breeding between pure parents and hybrids, which may keep a sector of the swarm relatively fit. Others—mostly second, third, and fourth generations, et cetera—reproduce poorly, so the overall population has less vigor. Remember, all offspring of hybrids are hybrids. None reascend the family tree to become pure cutthroat trout. These are early days. Future studies will look at other parameters of fitness, like age-class survival.

Standing now at Avalanche Lake, I watch the cutthroats break the water. Because of a steep gorge downstream, rainbows cannot swim to this rare oasis and dilute the gene pool. Pure natives find refuge here. I jot down in my notebook: Are these the only fish worth protecting?

The question is complicated and prompts other considerations. How much hybridization is permissible while still maintaining the evolutionary legacy of a species? Certainly less than a twenty percent mixture, Muhlfeld says, since this cuts fitness in half. Is it ten percent? Five percent? One percent or less? Both questions are important, and hotly debated among scientists and conservationists, as the legal status of hybrids is vital to the future of endangered species law. Species as diverse as the spotted owl and the red wolf have uncertain lineages. As it stands, however, there is no official policy on how to treat hybrids under the ESA.

A conservation genetics team led by Fred Allendorf of the University of Montana has looked closely at these issues, often using the westslope cutthroat trout as a model. They believe it is a mistake to rely on the morphology ("the looks") of the species for classifying pure from hybrid strands. In the case of the cutthroat, morphology can be misleading since even pure rainbows can share some traits—for example, the number of gill rakers—in common with the cutthroat. Genotyping is preferred, and techniques have improved in the past decade to make this possible on a large scale. While Allendorf thinks the threshold for each species should be considered on a case-by-case basis, he has championed the definition of a "pure" cutthroat to be less than one percent genetic mixing. Muhlfeld calls for zero. But for management purposes, which often require compromise, he calls for a standard of less than five percent. He recommends that only these "pure" populations of cutthroat be considered for listing under the ESA. The official range of the fish then may be less than two thousand miles—one-twelfth what USFWS now claims. The hybrid populations (in the other eleven-twelfths) pose a threat to the few remaining "pure" fish and, some say, these fish should be eliminated. Captured and killed. Purity, Muhlfeld asserts, is worth preserving. So is the evolutionary legacy of species and ecosystems. Climate change is making these tasks more difficult each day.

On Avalanche Lake, the cutthroats have fallen silent. Feeding time is over. The water surface is now smooth as glass ice, unbroken by fin or snout. I have a moment to look over the lake, thankful that the cutthroats have a refuge here. They are kings of the lake—the top fish, without peer. Occasionally growing to among the largest of all native, western trout species (over eighteen inches), it's possible the granddaddy of all cutthroats is here. A nice thought. King of kings.

I notice a flat stone—the size of a quarter—on the lakeshore, then another, and pick up a few. I haven't skipped stones in years but I have the basin to myself—me and the loon and the fish. With my left arm, I swing a wide horizontal arc and slide the pebble outward off my fingertips. Just for good measure, I put a little "english" in my wrist upon release. The stone skips off the water twice—*plink, plink*—then plummets. So I try again. And again. This repetition becomes a meditative act and allows me to step back for a few minutes to contemplate the mountains and their waters in time.

In the years before preserves and parks, cutthroats swam

freely along the three great drainages of Montana's northwest. The waters were cold, chilled by the ice of glaciers, by the snowpack, by the cool night air. The underwater life had adapted to the frigid conditions; all the aquatic residents had been here for thousands of years. They were alone and purebred. Exotic species had not trespassed into the native territory, nor had they been released by fishery managers. Humans walked into North America only recently, exploiting and finally modifying the waters to meet their own design.

Now, the climate is damaged, and the water is warming. The land is also not immune. Even for remote outposts like Glacier, the landscape has become frayed at the edges.

Like the creeks and rivers, Glacier's wilderness was originally preserved intact. Congress set aside the Park to place it outside the reach of blade and saw. Timbering was a direct threat—you could see it coming. By shutting out the lumber mills, this pocket of wilderness was saved. The Crown ecosystem was in its natural state here, surrounded by a buffer of forests and Great Plains. It was pure wilderness, untouched and uncompromised. And so it remained. Despite pressure from outsiders, no hunting or mining, no firewood gathering, no mountain bikes or snowmobiles have been allowed. While the surrounding forests were cut, while cities grew tall, Glacier National Park offered a refuge, where all the great herbivores and carnivores could roam, where fish could swim. But now the climate of cars and the cities blows across the peaks. The temperatures climb. Snow turns into rain. The water towers melt early and quickly. The streams run close to dry. Is it still wilderness or is it some sort of hybrid, mixing it up with the urban world? Has civilization and nature interbred? If so, how much hybridization can a wilderness endure?

As the stones skip across the water, the questions multiply in my head. I scan the waterfalls above and picture the exit stream from Sperry subsiding to a trickle. Fagre and Muhlfeld give Sperry less than twenty years. What will become of the cutthroat? Of the stonefly? Of Avalanche Lake?

My best stone bounces off the surface six times before settling to the bottom. Quitting time—while I'm ahead. I drop my handful of pebbles on the shore. They click on the rocks like crickets chirping. The last waves reach my toes, and I scan the smooth lake. Silence. Then, magically, fish break the surface. Ripples widen into circles. As soon as I stop tossing stones, the cutthroats start feeding again.

Fire and Ice

Over the next couple of weeks, I follow Fagre's team through the survey season from up close and afar. A little divine distance is helpful: I want to get the big picture. And I'm curious to explore more of the downstream consequences of glacier loss. Some of the legwork below the ice fields will be a solo journey. The impacts of climate change are numerous and scattered throughout the Park: wildlife, forests, and watersheds. To get to them, often well below the glaciers, requires stomping down an unbeaten path. I set out with a guide or on my own.

During August, the end of the month, I am early for a dawn rendezvous at the lab. I have a hike in mind, separate from the group, but I have a question for them. I stand in the lab's parking lot, hemmed in by trees. Overhead the stars are sharp but tattered like white bursts against a black target. Each star has its own signature, I realize. Each appears to sparkle on its own clock, distinct from the others. Some are brighter than their brethren. More intense, a few of these may have a shorter lifespan. That's the price of star power. At center, the constellation Orion—the great hunter—dominates the night sky. The Big Dipper sits above it, pointing to Polaris. For centuries, adventurers have oriented to that point to head north into interminable ice. Today, I will navigate in the same direction, only for a few specks of frozen water. I'm anxious about the day. Ice weighs heavy on my mind.

As it turns out, everyone has a different agenda today. Fagre is measuring a recent avalanche; Kevin will be counting tree rings nearby; Erich is running the trail to Blackfoot Glacier to check the snow cover; Lindsey is heading for Pitamakan Peak to survey wildflowers; and Lisa is digging through old pictures. Each is intent on evaluating climate change in a particular way. When they arrive, a light wash of yellow claims the eastern sky. The stars wink out.

I'm here to ask advice before climbing Swiftcurrent Mountain, which sports a fire lookout tower on top. I visited the larger of its two glaciers last week but kept to the foot. I hope to explore more of it, if time and safety allow. (In the mountains, distances and dangers are unpredictable.) Because of the hazards, I plan to ask Fagre's consent to trespass on the study site there. Technically, I already had transgressed, but forgiveness is easier to obtain than permission.

I think of the mountaineer's code—safety first—and begin there. I ask if there are crevasses on Swiftcurrent Glacier, lurking under the snow.

"Every glacier has the chance for a lethal crevasse," Fagre warns. "Be careful. I'd stay at the edge. You're free to roam the glacier—you won't hurt our survey—but avoid that headwall."

Tacit approval for my prior indiscretion. I'm good to go.

I motor off to rendezvous with my guide for the day—Sherry

Bateman. A former whitewater instructor, she is now a part-time mountain guide and mother of two. She has climbed Swiftcurrent Mountain many times and promises to get me close to the two glaciers that border the peak. "There's fire up there as well as ice," she announces. "A little piece of heaven, a little piece of hell." Sherry was one of the first to reconnoiter the burn zone below the "Swifty Lookout" after the devastating Trapper Fire of 2003.

Sherry holds the steering wheel of the van. We travel more than twenty miles along and above McDonald Creek toward the Loop, a 180-degree turn on the Going-to-the-Sun Road. The road, an engineering miracle, transects the Park west to east, crossing right over the Continental Divide. We will stop short of the crest at the Loop trailhead. On the way, we pass through a lodgepole pine forest. A lone bull moose stands in the creek. Farther along, Bird Woman Falls drains a cirque where a glacier once stood. Numerous avalanche tracks cross the road; one from 2009 took out a section of the parkway and blocked traffic for a couple of weeks. This is Fagre's destination for the day. His mission: To pin down the role of snow slides in mountain ecology. Beyond the avalanche, Sherry spots some fire damage in the creek valley and beyond. At that moment of the day, I don't yet grasp the parallels between fire and avalanche, nor their connection to climate change. The day has much in store.

The hike from the Loop parking lot to Swiftcurrent Pass is five miles of uphill trudge. But a prize awaits: Granite Park Chalet has candy bars for $1.25 a pop. At milepost five, I'd pay the chalet manager ten times that amount. Constructed in 1914, Granite Park Chalet has three stone-and-log buildings, sleeping and feeding up to forty hikers. Supplies are brought in by horseback. Staying there is one of the singular experiences in the Park. The views are stunning. I count five glaciers in the distance: Blackfoot and Jackson to the south and Rainbow, Two Ocean, and Vulture to the northwest. More lie up ahead.

Above the chalet, we pass through a series of alpine meadows, resplendent with wildflowers at the height of their song. Sherry is an expert in botany and announces each species with a little drumroll of her trekking poles. Yellow *Aster*. Purple *Anemone*. Yellow *Arnica*. Blue *Delphinium*. White *Heracleum*. A sampling of colors and the alphabet. Then I see red: scarlet paintbrush, though we both can't recall its Latin name.

We arrive at Swiftcurrent Pass by 11 A.M., after a climb of 3,200 feet (975 meters). This saddle on the Continental Divide is employed by man and beast to cross freely from the vicinity of Granite Park Chalet on the west to Many Glacier Hotel on the east. It is also a high alpine passage to the lookout and to Swiftcurrent Glacier, tucked just north of Grinnell. The eastern divide rolls outward in a series of twin peaks and canyons, valleys, and chain lakes—Swiftcurrent Lake, Lake Sherburne, Lower St. Mary Lake. The whole world is blue and green and brown. Like the first sight of a new frontier, the view makes you gasp in wonder. It hits you right in the chest like a fist.

I will return to this view once more over the summer, never bringing enough film to do it justice. Sherry is anxious to get the climb of Swifty's summit under way after a lunch break. In that twenty minutes, I can tell the story of my side trip to the glacier, itself, only six days before.

It was a brilliant day, the light sharp. From the pass, a mosaic of trees and meadows, I trailblazed to its southern shoulder. After a rise of only a couple of hundred feet, I came upon a long moraine, then the edge of Swiftcurrent Glacier, the gray ice and white snow blinding me for a moment. I slipped on my sunglasses and scanned the small glacier for signs of the coming ruin. Like a scout, I patrolled the ice at the margin, where the frozen water had pulled back, separating from the rock.

The Swiftcurrent ice field reminded me of a miniature Sperry: It's about one-fourth the size but has the same layout. From a steep headwall where snow accumulates and metamorphoses into ice, the glacier has a gentle slope and stops just short of a precipice. It's another hanging glacier. Waterfalls drain the glacier below, and their streams converge into Swiftcurrent Creek. The creek is home to the meltwater stonefly, just like the exit from Sperry. If the stonefly becomes listed as an endangered species, the Swift-current drainage may also be designated as "critical habitat," which may mean both glaciers require protection. Such a measure would have sweeping consequences for minding and mending climate change.

I tell Sherry some of these particulars, and though she is already well acquainted with the glaciers, she's hungry for more news—especially about Fagre's forecast. Like many others, she hopes for evidence that the ice fields will be okay. She has a stake in the outcome. Her trade—ecotourism—depends on the ice for mountain travel and whitewater. Sherry traversed the flank of Swiftcurrent Glacier once to climb Mt. Gould, which looms above us like a threatening fist.

Meanwhile, the Swiftcurrent cirque sits below like an empty palm. I describe the bowl to Sherry. This summer, half the expanse is bare rock, I say. She frowns. Then I share what I've learned from Fagre. A photograph from circa 1930 shows the ice coverage to be twice its present size of 50 acres (0.2 square kilometers)—about forty football fields. That's one-third the size of Grinnell.

Since 1966, Swiftcurrent has declined in area by about fifteen percent. The decay may be accelerating. Nearby North Swiftcurrent Glacier has lost more than thirty percent and is considered at 20 acres (0.08 square kilometers) to be under the threshold size of a true glacier. (Remember: The tipping point is twenty-five acres.) The main Swiftcurrent Glacier is one of the next in line for diminishment. If it loses another fifty percent of its area, it will be retired from the active roster.

The Centennial Glacier Retreat Report, released by Fagre in 2010, revealed other losses. Besides North Swiftcurrent, another ten glaciers over the twenty-five-acre threshold in 1966 dropped below that level by 2005, forty years later. (Gem Glacier was already too small to be a true glacier in 1966.) An ice mass is no longer considered a glacier at under twenty-five acres partly because it is unlikely to move at that size. The greatest loss was suffered by Harris Glacier, which lost 77.4 percent of its ice. In all, the twelve inactive glaciers (including Gem) lost nearly sixty percent of their combined area. Some glaciers lost more acreage than others, thanks to a south-facing orientation and more exposure to the sun. A number of remnants, or glacierets, are so small—like Clements, now a stagnant ice mass—that they aren't even counted in the tally.

Keeping track of the remaining glaciers and glacierets is Fagre's responsibility but, due to budget and staff limitations, he must pick and choose which ones to study intensely. Selecting them is akin to handicapping at the racetrack. Some will win a few more years; some will lose more swiftly. Besides Sperry, his annual benchmark glacier, Fagre monitors six "secondary" glaciers: Agassiz, Blackfoot/Jackson, Chaney, Stanton, Swiftcurrent, and Grinnell. These six are surveyed every other summer, weather permitting.

All tend to face north or northeast. Each receives the same treatment: margin profiling with GPS, snow-line estimation, and repeat photography. This trifecta excludes mass balance, which is solely reserved for Sperry. The week when Sperry was canvassed, the rest of the field took a rest. They were all covered by snow, making margin measurements difficult. Now, with a hot late August, melting is on the advance again. At least two of the secondary glaciers will be corralled through September.

The roundup began with Swiftcurrent, I explain to Sherry, as we huddle against a stiff wind, gulping down our sandwiches at the pass. Ten days ago—four days before my trespass—Kevin and Erich paced out the margins and recorded the coordinates with their GPS gear. Swiftcurrent had not been measured for three years, so the results would prove valuable.

Among the other secondaries, the margin statistics would also bring bad news: The yearly acreage losses are twenty percent higher in the last ten years than in the previous decade. Melting is speeding up. Rumor has it that Dan Fagre may be reworking his forecast.

As we pack up after lunch, Sherry tells me that the word on the trail is that Upper Grinnell Lake, right below Grinnell Glacier, has been coated with ice all summer. So, I realize, the odds for using the boats are still a long shot. Everyone will have to keep an eye out to see if and when the lake runs free. In any case, Fagre may want to wait until September first, when all melting subsides but before the next freeze. Then he can do a complete work-up on Grinnell.

A remote video camera, stationed on a cliff above Grinnell Glacier, would do the trick: catching the moment the lake melts free. As it happens, the second best thing sits above Grinnell and

Swiftcurrent. We will hike right past one of these still cameras, an "extreme ice survey recorder," on our ascent of Swiftcurrent Mountain.

The Extreme Ice Survey (EIS) is a global network of remote, time-lapse cameras positioned to document rapid glacier loss. Over forty cameras have been installed on twenty-two glaciers in Greenland, Nepal, Alaska, the Rocky Mountains, and elsewhere. Each camera records the glacier at least once an hour for every hour of daylight, totaling over four thousand images each year. Solar panels recharge the batteries. The only maintenance required is to change the memory card annually. Worldwide, the cameras must face winds up to 170 miles per hour, blizzards, torrential rains, and temperatures as low as minus 40 degrees Fahrenheit (minus 40 degrees C). The camera bolted to the cliff above Grinnell Glacier was knocked free by rockfall, two years in a row. The still images will be edited into video to disseminate evidence of global warming.

Kevin is scheduled to swap the memory card any day now. It will likely show my trespassing on one of its frames. But there is greater treasure.

One startling European photograph obtained with an EIS repeat camera, specially programmed for an annual shoot, shows three tributary glaciers in the Alps shrinking back from the trunk they once shared. The Leschaux Glacier (at the foot of the rock-clad Grande Jorasses), the Talefre Glacier, and the Mer de Glace (both under Mont Blanc's rock and snow) were all once joined in a super glacier but have retreated into their own sub-basins of gray ice. The shriveled arms resemble three amputees.

Throughout the Alps and Andes, whole glacier complexes—valley trunks and canyon arms—are vanishing, and since 1980 at an accelerated pace. In Peru, the melting of the Quelccaya Ice Cap and its outlet glaciers has already prompted water shortages in Cuzco, a city of 400,000 people, who have resorted to periodic water rationing. Perhaps most striking in Europe is the retreat of dozens of glacier tributaries on the Mont Blanc Massif (15,782 feet; 4,810 meters), the highest and broadest mountain in the Alps. Known to locals as "La Dame Blanche" (French for "The White Lady"), she dominates the landscape between Chamonix, France, and Courmayeur, Italy, her base larger than the size of Manhattan, her summit more than ten times the height of the Empire State Building. One of its largest ice fingers, the Glacier d'Argentière, flows down the mountain for 5.6 miles (9 kilometers) toward the town of Chamonix. (It resides one valley north of the Mer de Glace.) Since 1870, Argentière has receded 3,770 feet (1,150 meters), pulling back farther into its own valley. In 2005, a line of seracs collapsed on the foot, a crashing and crushing loss of over 400,000 cubic meters of ice. Such a collapse is an indication of the disintegration of the glacier. Afterward the toe of the glacier dangled by only a thin strip of ice. French scientists estimate that the stranded toe may disappear in twenty years.

As a boy, I dreamt of climbing Mont Blanc (and read all the books about her; my father told me of seeing her on his march across France during World War II), but it wasn't until I was twenty-four that I traveled to the continent to try my luck. In the meantime, I trained for her. At fourteen, I attended the Exum Mountaineering School in Moose, Wyoming, and climbed Grand Teton (13,776 feet; 4,199 meters), which has some challenging rock near the summit. Over the next few years, I prepared for the Alps by summiting Mt. Marcy (5,344 feet; 1,629 meters) in the Adirondacks and Mt. Washington (6,283 feet; 1,917 meters) in the Presidential Range, both in winter, and, finally, after an aborted attempt, climbing Mt. Rainier (14,410 feet; 4,392 meters) under a cobalt-blue August sky. In college, I honed my skills further, chewing the leather with the campus climbing team, but I still longed for the White Lady. Our perilous courtship would come soon enough.

On my first trip to Chamonix, in June 1980, my initial stop was the legendary Compagnie des Guides to check out announcements on the blackboard, and I was disappointed to discover the mountain had been shut down indefinitely due to blizzards and deep snows. Avalanches were imminent. If a team dared to trespass, it would be on its own—no rescues would be launched. In the guide hut, I stood next to another solo American climber, Roger Singer, who was also looking for a partner. Beneath the chalkboard, we commiserated and joined ranks. Both of us had five days before our return flights to the States and pledged to climb Mont Blanc within that window. We shook on it. The climb would be in secret; there would be no record of our attempt.

The next day was overcast, though not snowing. The tramways and cableways were not running, thanks to the storm of the previous days, so we hiked from Chamonix Valley (approx. 3,300 feet; 1,006 meters) along the tracks to the Nid d'Aigle ("Eagle's Nest") train station. From there, we trudged through waist-deep powder to the Tête Rousse Hut (10,390 feet; 3,167 meters) on the west side of the mountain. The steeper work lay up ahead, but we had made a smooth start. The break in the storm was welcome and our timing seemed auspicious: Perhaps the weather had lifted. I rubbed a good-luck stone in my pocket—as a wish and in thanks.

After a short rest, at 3 A.M., we drank some hot chocolate, strapped on crampons, tied on a rope, and headed with ice axes in hand for the Aiguille de Goûter (the "Goûter Needle"), a pinnacle of rock and legend. The needle was over 2,300 feet (700 meters) high and divided the Tête Rousse and Bionnassay Glaciers. A jagged buttress, the needle was composed of long vertical ribs of granite, which were threaded by alternating avalanche chutes. The biggest gulley was called the Grande Couloir and had witnessed its share of fatalities. The safer route was to hug one of the rock ribs on either side of the couloir, climbing the rock, not the snow.

But when Roger and I got to the face, we discovered three feet of snow on the ribs. Just the Grand Couloir was free of snow. Being so steep, it had already avalanched. It was down to bare ice.

As the stars gave way to the first hint of dawn, I was leading the first section of ice on the Grande Couloir. We changed positions, from leading to seconding, every rope length to relieve the front man from the exhausting effort of kicking and chopping steps in the hard ice. We climbed quickly, aware that the rising sun would warm the couloir, likely releasing rockfall at the top.

By 7 A.M. we were two-thirds of the way up the gully. The wind, nearly absent before, seemed to pick up, and ice crystals swirled around me and stung my face. Next, I heard a strange buzzing. It got louder. I feared the worst and threw my ice axe deep into the ice, leaned into the slope, and held on for dear life. Nothing happened. I looked down at Roger and he pointed behind me into the dawn light.

A helicopter was hovering at eye level about a hundred meters away. They were filming the madmen. We waved the pilot off—flipped him the bird, I think—and the helicopter dipped and disappeared. Later, we learned it was a camera crew from TF1, the premier French television station. Roger climbed up to me and, at that moment, either stimulated by the warming sun or the helicopter blades, the first pebbles rained down on us from above. Roger insisted we unrope, and I could see the logic behind his precaution: Speed was our only chance for safety, and ropework slows one down.

Without the rope, if one of us fell, the other climber might survive.

Just before we reached the headwall, a small rock hit me like a bullet in the left eye. My glacier glasses shattered. Miraculously, that was the limit of the damage. At 8 A.M., we surmounted the rocks of the ridge and stood atop the Aiguille de Goûter.

The Club Alpin Français had ingeniously placed another refuge—the Goûter Hut—at the top of the ridge. We still had 3,258 feet (993 meters) to the summit, so we settled into the sanctuary, to rest until midnight. The summit bid would be a snow ascent—not rock or ice—but mountaineering is always best before dawn, when the snow is firm.

Yet the next leg of the climb would be delayed. Storm clouds reconfigured to the west. By the time midnight came, winds over 100 miles per hour (166 kilometers per hour) pounded the hut. They blew heavily the whole next day. It was difficult to tell whether it was snowing or whether snow was drifting down from higher on the mountain. We had to wait. Two days later, the wind died enough to allow us to stand outside the hut. Time to dash for the summit.

The climbing was easy compared to the ice of the Grand Couloir. At 7 A.M. on day four, we passed over the Dôme de Goûter, a false summit, under a blue sky. Then, quite suddenly, the White Lady was up to her antics again. A fierce wind blew from the south over the Glacier du Dôme, and clouds appeared out of nowhere. Wisps of mist, like streamers, shot straight up overhead. By now, we were ascending a thin knife edge, the Bosses ridge. To keep balance, I straddled the razor with the right side of my body in Italy and the left side of me in France. I negotiated the ridge like compromising on a treaty. At 10:30 A.M. we saw the summit flash briefly in a break between clouds. Then there was total whiteness.

The wind was deafening. Roger and I huddled, yelling into each other's faces. I think Roger asked me if I wanted to turn back. I did. He read me wrong and pointed a thumb upward. I did the same.

We were going to make it.

We leaned into the blizzard. We began to crawl. Roger was ahead when the slope leveled off into a snow peak that ran east-west—the well-known orientation of the summit.

Roger put away his compass. He yelled, "This is it!" I checked my compass and agreed. The view was nil. Roger screamed, "Let's head down." I nodded my head, but before descending I buried my good-luck stone in the summit snow. My father had carried the talisman across Europe during World War II.

We would spend another night on the mountain—in the Valot Hut—and then rush down to Chamonix and our flights home. On the descent, two French climbers asked if we were "the crazy Americans." They had seen us on the television, which had played apparently like an early obit.

The White Lady had been kind. Ever since then, I have always considered mountains to be feminine and refer to her and my other mistresses the way captains do their boats. "She has wonderful lines," I have been known to say. Could be a peak or a ship. Or a woman, of course. In Japan's high country, the locals call snow "yang" (male) and ice and water "yin" (female). A glacier is born male but evolves into a river of femininity. And at the top of it all, Everest is called Chomolunga, the "mother goddess of the earth." To be sure, the gender of summits is open to dialect and interpretation. But I prefer the French vernacular. In any case, the White Lady is why I love mountains—because she and her kind play so hard to get.

The sun hangs just past noon, as Sherry and I begin to motivate. We stand and stretch. On a whim, we split our third walnut brownie before climbing the fire lookout mountain. But on the last bite, Sherry has a different idea.

She says, "Let's take another look at Swiftcurrent Glacier." My curiosity is up. "Ice changes a little all summer long." Sherry reaches down and hefts her pack to her shoulders, all in one motion. "If we hurry," she says, "we'll have plenty of time afterwards to climb Swifty, too."

I agree. The final call on the itinerary is always up to the guide, but my wish has been fulfilled. Swifty hovers above, behind my ear, like a patient teacher. Whatever lessons she has to offer will have to wait.

We trace my old footsteps—from the pass, over the shoulder, around the moraine. We reach the ice in forty minutes. Recalling Fagre's advice, we keep to the edge. On our approach, a jet-black raven lifts from a white depression at the lip of the glacier and flies away. An ebony bird with snow on its talons.

Glaciers are once again a study in contrast.

We are now standing at the foot of Swiftcurrent Glacier, poised at the lip of a smiling crevasse. The inner recesses of the gaping hole retreat into blackness, but the six-foot-wide mouth flashes blue ice—the pastel color of today's sky. The clear ice absorbs the majority of the red-and-orange spectrum of the afternoon sunlight, reflecting only blue. At the edge of the crevasse, on a small shelf just within the lower lip, I find a pocket of snow. Shielded from the full onslaught of the sun's rays, the snow has survived from last winter or a late snowfall in June. In contrast to the blue ice, the snow is blinding white. The combination reminds me of blue-and-white porcelain on a white tablecloth.

Walking gingerly, I venture along the lip of the crevasse and peer down, past the snow-covered ledge, to the black bottom. Sherry tells me crevasses tend to be 100 to 140 feet deep, on average (30 to 43 meters). I take her word for it and don't lean too far. Every glacier I have visited so far has presented at least one crevasse as humbling as this one. I've jumped across a few in my day but I'm always respectful. I retreat to the fringes for a few moments of safety. Sherry wanders off to find more ravens. The cirque keeps time on many clocks, not all slow ones: I watch two birds take wing in a zigzag like black lightning. The pair calls out "tok-tok."

Not only is the raven's speed on takeoff remarkable but its range is formidable. This pair may be visiting from a lakeshore far below. Hearing their call takes me back to another mountain— Denali in Alaska—where ravens regularly fly twenty or more miles to scavenge on climbers' caches, "hidden" in the Muldrow Glacier. The birds recognize the shovel marks in the deep snow. On my climb there in 1976, ravens dug up a week's worth of our rations, gorged themselves, and endangered the completion of our ascent. We never saw them—just their footprints next to our empty food bags on the ice. And their bold hunting calls—tok-tok-tok—in the mist. Mischievous and greedy, yes. Also, smart and wild. I'm happy to see them today.

Far above us, at the headwall of the glacier, snow has accumulated, too. This is the snow that, upon turning to ice, would typically add to the growth of the glacier. But the ice, on balance, is retreating. Even with recent record snowfalls, the mass balance is simply overwhelmed by melting. The headwall snow is consolidating, compacting to a tighter density, but it is not yet firn, not yet ice. This is snow's historic destiny: impermanence.

From the crevasse shelf, I pick up a dab of snow on my fingertip and inspect it. Out of my pocket comes a magnifying loupe, brought along to view the rocks. The loupe has three plastic discs for viewing at different magnifications. I choose the boldest. The snow crystals from the crevasse do not resemble snowflakes; they are slightly rounded, lacking dendrites and frills. They look more like muted sea stars than lace. This is old snow, consolidating and bonding ("sintering") with its neighbors. It has begun its journey toward ice.

Still, the snow crystals are hardly bigger than a grain of salt. It's tempting to calculate how many snowflakes create this crevasse, that headwall, or the entire glacier. Untold zillions, I believe. The zeros are lost to the calculator and to the brain. But I try. A cubic foot of snow may contain up to ten million snowflakes, but once it (and more snow) compresses and mutates to a cubic foot of ice, the count is likely in the billions. A glacier may contain tens of millions of cubic feet of ice or more. All glaciers on Earth—nearly 400,000 of them—descend from snowflakes. The mind reels. Too many zeros. I give up. I return my gaze to the crystals.

All snow crystals are hexagonal, even these fading stars. This six-sided symmetry mimics the structure of the water molecule itself. Emerging from the hexagonal template are perhaps an infinite number of incarnations. The oft-cited claim that no two snowflakes are alike may be true. No one has disproved it. As far as we know, snow crystals are as unique as fingerprints. The shape they take depends on the amount of water vapor and the temperature of the ambient air. For example, very cold conditions (temperatures between -13 degrees F and -58 degrees F; -25 degrees C to -50 degrees C) create simple, hollow columns. Meanwhile, high humidity nurtures lacelike growth at the corners and edges of ornate stars.

It all begins with a speck of dust (or salt, smoke, or another microscopic particle), lingering in a cloud. This singularity forms the nucleus of a snow crystal—the catalyst. Nearby water evaporates off a supercooled droplet, and the vapor condenses around the nucleus, instantly freezing. The water leaps from vapor to a frozen crystal, without ever passing through liquid. The snow crystals mature into a variety of shapes, which have been classified into seven categories: hexagonal flat plates, stellar crystals, columns, thin needles, spatial dendrites (with nonplanar branches and arms), capped columns, and irregular crystals (a grab bag of uncatalogued shapes). The classic star-shaped crystal—either a stellar or a dendrite—is not the most common. Look for it at intermediate temperatures between 3 degrees F (-16 degrees C) and 10 degrees F (-12 degrees C).

Let's return to the cloud. For crystals to fall as snow, they must grow to a large enough size to overcome updrafts and for gravity to do its work. The simplest way for this to happen is for crystals to collide, forming the clusters we know as snowflakes. The optimal conditions for snowflake formation—several snow crystals sticking together—is when the air is warm and the crystals are wet, typically from high humidity. During the lifetime of the

Earth, it has been estimated 1.0 x 10^{35} snowflakes have fallen, which would cover the entire surface, oceans included, to a depth of fifty miles—if there were no melting.

I blow the snow off my fingertip. A slight wind catches the crystals. They swirl. They drift in an elliptical orbit, and finally collapse into the broad crevasse. I watch them descend. Against the black curtain they resemble stardust—they glitter as they fall. Each snowflake has a history: Some melt right away, some evaporate, some are destined for ice. Billions of water molecules traveled far to assemble here. They carry dust and pollen from around the world, brought here by the wind. Some from Africa; some from the Thames; some from the Amazon; some from Mont Blanc. Any snowball or cup of water contains molecules that quenched Cleopatra's thirst. They carry the wellspring of life: water to quench the entire world. Water is the ultimate frequent flyer, the ultimate recycler. I will see these snow crystals, or their constituents, again.

Just now, a small snowslide careens down the headwall. I'd call it a "sluff," a harmless minor slide. The sluff tumbles layer by layer. More than a dozen winter snowfalls surface, then are buried again, like a chef turning over a vanilla cake.

While a glacier recycles water into many guises, an avalanche track has many incarnations, too. Avalanches, it turns out, are not haphazard, random events but cyclical phenomena tied to the climate. Like a glacier, the life of an avalanche begins when a snowflake falls. In Montana, this commences with the first relatively warm snowstorm of the autumn, and that initial wet layer can often be the catalyst for an avalanche. This is likely when early snowfalls are followed by a few weeks of clear, cold skies. The thin snow layer is sandwiched between the warm ground (radiating summer heat) and the cold air on top. This temperature gradient promotes vapor flow through the snow, creating a frost called "depth hoar" that is weak and unstable. When more snow accumulates on this weak base, the slope is primed for a slide—a quick burst of snow. The life of an avalanche is short but violent.

On all mountain slopes, the layering, or stratigraphy, of the snow determines the likelihood of an avalanche. When a slide's anatomy is inspected postmortem, for forensic clues, invariably a weak layer is discovered underfoot. The snowpack's weakness is the avalanche's strength. The deeper the weakness—and the more inches of successive snowfall above—the more powerful the slide.

Over time, changes take place in the snow layers as they consolidate and become more dense, in some cases making avalanches more likely. At first, a typical blanket of new snow is eighty to ninety percent empty space. How quickly the snow begins to compress is controlled largely by the shape of the snow crystals. Needles, for example, pack more closely than stellars. They produce wind crust on ridgetops. This crust often forms during a period of fierce winds between snowstorms. As the crystals of any given type bond and change, they become more rounded and settle into a tighter mass. Because each layer represents a single snowstorm that fell at a particular temperature and humidity, each layer has a signature crystal or two—a thin needle, for example—that pack together in a similar way.

The weak link is not always a buried snow layer, however. In addition to wind crusts, a melt-freeze crust may appear between storms, adding a slick glaze to the top surface of the last snow section. The next snowfall after that encounters little friction, as if settling on a sheet of wax paper. On any significant incline, the new passenger is poised for a slide.

Another hazard, Sherry says, returning from her jaunt, is "surface hoar," the winter equivalent of dew which can precipitate on a snow surface on clear, calm nights. This is the sparkling layer of feathery flakes that emits a pleasant hissing sound for the skier. Innocent enough on the surface but, when buried, surface hoar can be deadly. On a steep slope, it can act like a rack of ball bearings to the new snowfall above. A sheet of snow, six feet thick or more, can careen down a mountain at up to two hundred miles per hour, taking out trees, buildings, and skiers in its path.

Generally speaking, there are two basic kinds of snowslides: the loose-snow type and the slab avalanche. Both have many variations, but the two general categories are distinct. A loose-snow avalanche consists of cohesionless snow, like powder, and initiates at a single point, from where it spreads out in an inverted V-shape down the slope. Meanwhile, a slab avalanche—comprised of either dry or wet snow—requires a cohesive snow layer that is poorly anchored to a weak layer below. The slab release commences with a broad area and, as the section breaks free, can add up to an enormous amount of snow. Surface hoar or depth hoar hidden beneath a top layer or two of fresh snow can prime the slope for a major slide. Those ball bearings are more efficient the steeper the slope.

Besides the prerequisite weak layer of a slab avalanche, other factors—terrain, snow depth, wind, rain, temperature (both ambient and the temperature gradient within the snowpack)—play an important role in avalanche probability and power. Snowfall is also a key ingredient. Over eighty percent of all avalanches fall during or immediately after a storm. And the crystals are key: Plates, stellars, and dendrites have little cohesion and form low-density powder; columns and fine needles pack tightly. The first group could spawn a loose-snow avalanche; the second may form a cohesive band on a slab. All these factors are evaluated in the formulations of the snow avalanche forecaster.

The time-lapse cameras of the Extreme Ice Survey have recorded snow avalanches on the ridges above glaciers. They have also captured the moment of another type of slide: the ice avalanche. The ice variety, sloughing off a glacier itself, contains some snow and, of course, is the product of the original snow that fell a year (or fifty or more) before. But it is predominantly ice. Typically, ice avalanches sever a chunk off the face of a glacier. Agassiz Glacier at the north end of the Park is known for its icefalls. In Glacier Bay, Alaska, and elsewhere in the far north and far south, glaciers often "calve" right into the sea, creating icebergs with a splash. My last encounter with falling ice was closer to home—on Mt. Rainier in Washington State.

The year was 1985. On our third climb of Mt. Rainier, my rope partner, Jake Stout, and I decided to take a new path to the summit. Rainier had already begged our patience during the harrowing rescue on the northeast side of the mountain, where we played a part in extracting three climbers from a deep crevasse. More recently, in 1981, eleven climbers were crushed to death on Rainier's Ingraham Glacier when a monstrous chunk of blue ice broke off above them—the worst avalanche accident in U.S. history. So we were cautious, taking our time.

Bypassing traditional southern routes, like Disappointment Cleaver via Camp Muir, we headed west from Paradise Inn across the Nisqually Glacier to the Kautz Route before dawn. This path offered three spectacular variations up or beside the legendary and treacherous icefalls above Camp Hazard, a bivouac site at about 11,300 feet (3,444 meters). We were never to get close enough to choose between the three. But that was our salvation. We were running late.

Just after dawn, a shaft of light peeked behind Little Tahoma, the false summit on Rainier's southeast side, and hit the Kautz Ice Cliff above us. The 500-foot, towering ice face just needed that little nudge. The last holdfast melted and the whole wall came crushing down. From our perch a few hundred yards away, we watched the ice rip loose and shred into powder. Next we heard a cannon shot. A deafening roar reached over us, like freight cars crashing. By our estimate, an ice block 1,200 feet long, 300 feet deep, and 500 feet high rolled off the precipice—the volume of 45,000 school buses. It landed at our feet. Fortunately, we stood on a rock outcrop above the avalanche track and only suffered a makeshift "ice storm"—the spindrift of the airborne debris. If we had been fifteen minutes earlier on our climb—that is, farther along the route and on schedule—we would have been pulverized. I've been a confirmed procrastinator ever since.

While ice avalanches like this are awesome spectacles in the Pacific Northwest and Alaska, powder and slab avalanches are more common in Glacier National Park. There are simply fewer glaciers. Tens of thousands of snow avalanches rip through the backcountry of Glacier each year, with the combined destructive force of a tsunami.

The deadliest avalanche in Glacier's history—likely a slab slide prompted by depth hoar—happened on Mt. Cleveland, the Park's tallest peak at 10,466 feet (3,190 meters), in December 1969. Five young men, all under twenty-two, attempted to scale the heretofore unclimbed North Face. Weather or route conditions forced them to switch to the West Face, a move that confounded rescue teams. In any case, an avalanche caught them in mid-ascent. All five were swept down the face and killed.

Here on Swiftcurrent Glacier, the avalanche risk appears modest. The snow slopes above the glacier are not steep. The snow is not fresh; it is not loose; it is settling every day. Sherry and I are relaxed about it. Still, it's often impossible to tell *exactly* where a slide might be triggered. One of the biggest avalanches to descend past Going-to-the-Sun Road happened with no warning. The road is susceptible to snowslides: Over the years, two men have been killed by slides while plowing the road. But no one had seen an avalanche on this scale in perhaps fifty years. Mountains invite us to humility.

The Little Granite Avalanche, as it's now called, was misnamed. It wasn't rock—it was snow—and it wasn't little. It was colossal. The destructive slide originated near the crest of the Garden Wall in early January 2009, after a winter rainstorm, and plowed down 3,600 vertical feet (1,097 meters) of mountainside through five ecological zones ranging from alpine tundra to riparian forest. Hundreds of trees, up to forty inches in diameter, were uprooted and carried to the runout zone, which terminated near Upper McDonald Creek. The entire 1.8 mile (3 kilometer) path gouged the landscape and littered the roadway with a massive pile of trees, rocks, and snow that took several days to clear. The lateral track and roadside debris are on display each time I drive by. We saw them this morning. By all accounts, Little Granite was a Class 5 avalanche, the highest ranking on the U.S. scale.

Midwinter rain, an artifact of global warming and local warming in Montana, may have been one of the triggers that fueled the snowslide. Rain adds weight. An inch of water is the equivalent to ten to twelve inches of average snow. It also compromises the structure of the snow cover. This process is similar to meltwater's action. In the spring, as the sun softens the snow, meltwater percolates downward, dissolving the bonds between snow grains, weakening the snowpack. The strength of the uppermost layers crumbles, and any weak layer becomes even weaker. Meanwhile, the water runs until it hits an impervious layer—wind crust or depth hoar, for example—where it fans out, lubricating that weak layer, priming it for a slide—one of the reasons avalanches are more common in the spring. Rain has the same effect: It adds weight without adding any internal strength. In fact, it lessens cohesion. The feathery skeleton of the snow crystals simply dissolves, and the slope becomes less stable.

These conditions add up to what is known as a "wet slab avalanche," where a weak, wet layer breaks free and careens down the slope. At first, Fagre was suspicious that Little Granite was a wet slab event, but more recently his staff has suggested it may have been a dry slab formation since the avalanche originated above the rain zone, amidst dry snow. There were no witnesses, so pointing to an impetus is pure conjecture. However, what happened next is clear. The downhill course of the slide is visible for all to see. That slope had excessive loading in the form of recent snow, followed by rain halfway down—the coup de grace, in Fagre's words. As the slab rushed down it entrained the wet snow from the rain. Under these conditions, a huge, destructive avalanche was nearly inevitable, regardless of its origins.

Dan Fagre and Erich Peitzsch, his graduate student at the time, were among the first to arrive at the scene. They skied from Lake McDonald along the snow-packed road. When they reached the avalanche track, it looked as if a train full of timber had run an intersection, dumping a load thirty feet high at the crossing. They began to measure the track. The pair had been studying avalanches for a few seasons—to understand their role ecologically and to help the Park with forecasting big slides. The Little Granite track would become a living laboratory for their investigations. The first discovery was the shearing of 150-year-old conifers in the avalanche's path. This leveling of the forest would likely cause a shift in the dominant vegetation along the slide's trajectory. Within a year or two, aspen and alder seedlings, as well as early herbaceous colonizers, would invade the path. Monitoring the plant recovery in the wake of the 2009 avalanche would require a multi-year effort.

"Often, when you think you're proving the obvious, you find out you didn't know what you were talking about after all," says Fagre. So there will be no jumping to conclusions about the Little Granite recovery, but successional patterns elsewhere offer some clues.

Like other large natural disturbances, such as forest fires and floods, avalanches and their tracks sponsor new wildlife habitat, plant diversity, and sediment transport within a watershed. The ecological effects of a major slide can persist for decades, even centuries. Little Granite, for example, followed a track that had witnessed cyclical events perhaps for millennia, but the recent slide eclipsed the others, expanding the chute by thirty percent. That widened perimeter will likely be a long time recovering. In the meantime, if trends elsewhere are replicated here, the Little Granite chute and adjacent tracks will add to the heterogeneity of the mountain landscape. Forests will grow to alternate with meadows, like the parallel runs of a summer ski bowl.

Like Park fires, avalanches instill an ecological balance, an equilibrium in mountain habitats. Disturbance maintains complexity in species and niches; complexity nurtures stability. Once an avalanche removes forest cover, its destructive phase ends. From there, the linear track offers a template for the colonization of the raw ground by pioneer plants. Within a few seasons, meadows and shrublands paint the track with green. These nurseries can quickly become a mosaic, full of edges and contours, and in this way, become vital habitat for wildlife. Grizzly bears forage along avalanche paths for berries and for thawing carcasses of animals killed by the original slide or subsequent sluffs.

While hastening biodiversity, the alternating slides also protect adjacent forest by acting as natural firebreaks. No trees, no fuel. Adjoining mature forests are protected from subsequent infernos racing up the watershed. At the same time, avalanches nurture heterogeneity directly in their paths, which curbs disease outbreaks in trees by keeping a mix of species.

So, avalanches are a combination of destructive and regenerative forces. They are wild beasts. But in the wake of advances in climate and avalanche science, if not tamed, they at least are becoming more predictable. Dan Fagre and his team are helping to correct a misconception: Avalanches are not random environmental hazards. They are cyclical disturbances and, like fire, are forgers of the landscape.

Four years ago, Blase Reardon and Dan Fagre examined the historical record of avalanches along a different path in the Park. Over nine decades, rangers had recorded a dozen or so slides on that one track. But was that a full assessment? Just to make sure the rangers had been thorough, Reardon and Fagre cut 109 tree cross-sections on the same chute to record physical evidence of the slides. Trees that are injured by avalanches respond by growing scars, reaction wood, and other anomalies, which can be dated by their proximity to tree rings. The combined Park and tree-ring records yielded twenty-seven avalanche years out of the ninety-four-year period between 1910 and 2003. The rangers' historic record had underestimated the avalanche years by half. The tree rings spoke the truth: An avalanche occurred, on average, every three and a half years.

The avalanches seemed predictable but the timing impulse was uncertain until Reardon and Fagre started digging up other historical records. First, the nearby weather stations revealed temperature and precipitation data suggesting a pattern. It turned out that the avalanche years were correlated with excessive snowfall (in terms of snow-water equivalent) and rainfall. Wet snow and rain-on-snow events were prompting snowslides. More surprising, periods of the most frequent avalanches corresponded with the negative phase of the Pacific Decadal Oscillation. This is not a definitive correlation—they need a longer record of activity, say a hundred years—but it suggests years of cool, wet weather may be more likely to spawn avalanches in Glacier National Park.

These connections have profound implications for both avalanche forecasting and predictions of climate-avalanche relationships far into the future. Surveying and monitoring of avalanches in Glacier National Park are now running on a parallel course with glacier research. Fagre and his team are breaking new ground: The fledgling science of avalanche ecology has just been born.

Already, Fagre and Peitzsch have cut 130 cross-sections from dead trunks at Little Granite to inspect tree rings and avalanche scars. By calculating how frequently slides have come down the track in the past century, they hope to foretell when the next major cascade will fall. Avalanche modeling will also help. To shape a computer simulation, more data will be needed: temperature, solar radiation, and water content of the snow. Water collectors will be placed strategically above chutes to record when water is circulating though the snowpack. Such forecasting tools are now readily available—at a price. Fagre also likes historical records. He hopes to examine aerial photographs from 1945 to 1968 to see how frequently avalanches happened elsewhere in the Park. What is past is prologue, he likes to say. By pinning down previous conditions, he will be able to extrapolate into the future. "Then perhaps we can tell," he says, "when the next dangerous slides will come down the mountain."

This morning at the cabin, I asked Fagre what global warming will mean for avalanches. Will avalanches become rarer, as glaciers have, or will they proliferate in the wake of more heat and rain?

"My personal prediction," he said, "is that, in the short run, say over the next twenty years, we'll have more avalanches. I think they'll mostly be wet snow events, since winter rain and warmer temperatures will become more common. Rain and melting snow are such a big trigger. But, eventually, the reverse will happen— we'll get fewer slides. We will continue to lose our snowpack, thanks to melting in winter, so the snow line will recede higher into the mountains. There just won't be enough snow to create many avalanches."

Fagre foresees a stratified scenario: "We could see avalanches at higher elevations," he said, "but we won't have many reaching the valley floor, and those lower ones right now are the slides that have the greatest ecological effect. The high-elevation slides simply run over alpine tundra and rocks. The bottom avalanches rip out—and replenish—forests."

With broad strokes, Fagre painted a picture of how the mountains will look fifty years from now. "The long-term effects of losing the avalanche cycle will change the landscape. For starters, the current avalanche tracks will fill in with trees. All these trees will use water, so stream volume could reduce. And we'll already have less streamflow because of the shrinking snowpack and glaciers. Without the ecotones—the edges—we won't have the same wildlife habitat. This is the biggest concern: We'll have less biodiversity. The montane zone will become homogeneous. And what will happen next? Fires will break out and run for miles. We won't have those natural fuel breaks that the avalanche tracks give us now.

"For certain, we'll have *some* avalanches as long as we have snow, but they won't have a crucial ecological role like they do now." He caught his breath. "You know, I hope I'm wrong."

If Fagre is right, then the wet slab is the path of the future: a high-altitude, rain-influenced slide. "Rain is really heavy," Fagre emphasized. "It's like putting a steel plate on top of the snowpack. It compresses the snow, crushes the weak layer, and the whole thing slides. I think anytime we've had a rain-on-snow event in Glacier, we've had avalanches. That's global warming. I can't say any one event—like Little Granite—was caused by climate change, but thousands of wet snow avalanches elsewhere add up to a trend." Fagre punctuated his sentence with three nods of his head. At this moment, I realized hundreds of lesser avalanches probably fell the same day Little Granite did, right after the rain. Multiply this by the ever-increasing rain days in winter, and one can see how plentiful and powerful those slides become.

.   .   .

Standing watch over my crevasse full of aging snowflakes, I try counting the ice crystals in the glacier again. It's hard to resist such a challenge. But I lose the threads. I'd just as soon count the stars. My attention is diverted: A trickle of snow rolls down the sluff track on the opposite ridge. It barely touches the ice. Sherry rolls her eyes. She points at the modest runout, which serves to underscore the majesty of these once great ice fields, where ice and snow avalanches formerly ruled the day. Dying glaciers like Swiftcurrent and its kind are still prone to icefall as their cornices and seracs give way. And slides above Sperry Glacier contribute raw snow to its conveyor, the assembly line in its manufacture of ice. But now, scanning this diminished ice field—Swiftcurrent—I see that it is not on the verge of an avalanche, or even a surge, but imploding, withdrawing into itself. Soon its ice mass will be still, too small, too shallow to move. In its twilight, the glacier will steadily recede, retracting into its deepest frozen reserves until, in a final gasp, the last frozen water sublimates or melts into the land.

By now, it's mid-afternoon, and we have a three-hour hike down to the Loop, our parking area on the Going-to-the-Sun Road. But we take the time—two hours—to climb the mountain. Once off the glacier, the temperature climbs at least twenty degrees, pushing the 80-degree mark (27 degrees C)—hot for the Park. And it's dry. The summer drought has persisted since June. No rain has fallen in three weeks. Sherry jokes that all the rain fell last winter and now there's none left.

We circle Swiftcurrent Mountain, a mound of scree and switchbacks. Above us the whole way, the lookout tower peers down from the summit (8,436 feet; 2,571 meters). The lookout stands twenty-five feet above the peak, a stone masonry pedestal with a frame-and-glass cabin on top. This tower was one of the command posts for the nearby Trapper Fire in July 2003. The keeper, Christine Baker, hung tight while the blaze danced around her. That summer, a strong drought year, there were twenty-five fires in the Park (twice the typical number), burning 146,500 acres, or thirteen percent of the land. The National Park Service estimates that nearly one-third of all burnable vegetation—trees, saplings, shrubs, underbrush—went up in smoke that summer. The Trapper Fire was responsible for over twelve percent of that acreage. Lower stream flow in that sector from diminished snowpack was part of the problem.

Swiftcurrent Lookout was built in 1936. Since then, it has witnessed many fires, including the Heavens Peak Fire in its inaugural year that burned over Swiftcurrent Pass, torching 14,142 acres. But no year has been as devastating as 2003, which holds the record for the past century in acres lost. (The previous leader was 1910, known throughout the West as a profligate fire season.) Besides the Trapper and its cohorts in 2003, the past decade has seen another major blaze: In 2006, the Red Eagle Fire incinerated 34,203 acres, nearly half of that on the Blackfeet Reservation to the east. The increasing frequency of major fires has alarmed the National Park Service. Lookout towers like Swiftcurrent are more important than ever.

When lightning struck the Trapper Plateau, just northwest of the Loop, on July 18, 2003, conditions were ripe for a fire. It had not rained in three-and-a-half weeks, and the underbrush, fanned by a brisk wind, was bone-dry. The mean daily temperature was 68.7 degrees F (20 degrees C), with noontime heat spiking into the mid-eighties. At an elevation of 6,200 feet (1,890 meters), the plateau gets tons of snow but the snowpack was lower that year, about eighty percent of normal. The forest had been snow-free for three or four weeks. A Canadian plane spotted the lightning strike first: It ignited a spruce and alder grove.

By day two, it was up to fifteen acres.

The Trapper Fire burned for seven weeks, way into September, but the first ten days were the most violent. On day three, the NPS sent in a light team of hotshots to quell the blaze. The fire had passed the 3,600-acre mark and was not slowing down. The Trapper Plateau sits between the Livingston Mountain Range and the Lewis Mountain Range, where the southwest prevailing winds can race up the canyons. At first, the Trapper Fire ran down Upper McDonald Creek but quickly switched northeast, racing up Mineral Creek toward the Loop. Every spruce, birch, and alder tree in its path turned to flames.

Despite the burden of several more fires that broke out in the Park that week (the larger Robert Fire was one), the NPS employed a bigger team at Trapper. They cleared brush, dug trenches, and prayed. The fire was fast. "Nothing you could outrun," says Dennis Divoky, fire ecologist for the NPS. The only thing breaking its stride was an avalanche chute or two. On Day Nine, the fire made a run for the Loop and Granite Park Chalet, where forty-one people huddled, unable to evacuate: The air was clogged with smoke; visibility was poor; trails were unsafe. The Park Service received a Mayday call from the chalet. Christine Baker, the lookout on Swiftcurrent Mountain, persuaded the backpackers to remain calm. They were told to wait a day, then walk out—in the opposite direction of the fire—to Logan Pass. Nevertheless, four hikers did panic and bolted down the closed trail, in the direction of the fire, toward their cars at the Loop. For two days, they were feared lost. As it happened, they had a flash of good judgment on the trail, did an about-face, and marched out behind the others to Logan Pass.

Rain clouds rolled into view in September, and drenched the forest. The fire petered out but smoldered for weeks. The crews had helped contain it, suppressing the flames in a modest way. But, for the most part, the conflagration burned out on its own. When it was declared dead on September eleventh, the Trapper Fire had incinerated 18,702 acres—a tract just larger than the size of a small town.

The basic ingredients of a forest fire are dry fuel and a spark. Wind provides the catalyst for spreading the flames by drying the fuel further and by replenishing the forest floor and canopy with oxygen. Glacier National Park had been in a dry period for over a decade, with July rainfall accumulating less than an inch for seven of the past ten years. In 2003, the summer of Trapper, Robert, Bunyan, Middle Fork, Wolf Gun, and twenty other devastating Park fires, July had 0.04 inches of rain; August only double that amount. In the years ahead, climate change will likely usher in even more extreme weather. Whether it's winter rain or summer drought, the mountain ecosystem will feel the impact.

"We've had more acreage burn than ever before," says Dennis Divoky, "at least since the Park was founded in 1910. Some of the damage comes from greater drought, some from us suppressing the fires less." The Robert Fire, the largest blaze of that 2003 summer, hit close to Divoky's home: It threatened Park headquarters, Lake McDonald Lodge, and the entire town of West Glacier. All were spared by inches. Like Robert and Trapper, fifty percent of the past century's major fires have ignited in the past two decades in the Park.

Individually, these fires are not a catastrophe. We now realize fires are an important dimension of the mountain zone and valleys. These ecosystems evolved, reaching their potential under the influence of fire. Periodically, the forests are rejuvenated by successive waves of flame, thereby restoring the diversity of the ridges, slopes, and bottomlands. Many forest species are now adapted to fire, even depend on it. Lodgepole pine, for example, are seritonous, requiring extreme heat to open cones, to drop their seeds. Fire only becomes a crisis when it's a runaway blaze or a series of cataclysmic events, incinerating the landscape, like what happened in 2003 and 2006.

The first half of the Swifty trail takes us through burned-over subalpine fir from the Trapper Fire. Nine years and the forest is still a stand of charred sticks. Shortly, though, above tree line the trail switchbacks through bands of exposed limestone. Flakes of red argillite litter the path. High alpine plants, such as pink moss campion, blue speedwell, yellow stonecrop, and blue sky pilot, grow between talus and plates of stone. We gain the summit in a little over an hour, where we are treated to a rare panoramic view. Not only is the resident fire ranger able to see half the Park unimpeded but the count of glaciers is unsurpassed. I add up nine, ranging from Swiftcurrent and North Swiftcurrent (a remnant) to Sperry and Blackfoot and Gem. Only Grinnell, of those familiar to me, is hidden from view—in her case, behind a flank of Mt. Grinnell. All the others are up close and palpable.

The only alpine experience comparable to a great summit is the ease of the downhill leg, which can be just as exhilarating. On our descent, we watch our raven crevasse on the distant Swiftcurrent Glacier the whole way down. The gray-and-white expanse of the glacier, below us, suggests a frozen lake, the ice exposed in luminous patches. Not a large lake anymore. Maybe just a pond.

Several hikers join us for the hike from Granite Park Chalet down to the Loop. In bear country, it's nice to have company. And the noise. This path takes us right through the heart of the old Trapper Fire. Only minutes below the chalet we come upon the scorched earth. Charred spruce trunks stand in rows and groups, limbless and black, their branches burned to a nub. Elsewhere, patches of ebony intermix with pewter stumps like a grizzled beard. In the distance, on the opposite ridge, are broad bands of gray trees—also limbless—just sticks resembling thousands of fletchless arrows shot into a battlefield from above. There are few new saplings, just an occasional aspen, maple, or white pine. Divoky has told me that the fire burned so hot that it incinerated the "seed bank" beneath the soil and any cones aloft in the canopy. Lodgepole pines had not been common beforehand on the Trapper Plateau, so they haven't regenerated here. (Near Lake McDonald, the lodgepole has recolonized over the Robert Fire, however.) Trapper's spruce-hemlock community will be decades coming back.

Nevertheless, while the midstory and canopy are dead, around a bend in the trail appears a green forest floor with a cascade of wildflowers flowing down the mountainside. Clumps of monkey-flower with bright yellow blossoms nestle up against the white flowerheads of pearly everlasting. Nearby, goldenrod is in bloom. Above this, at knee height, are pink bouquets of fireweed, one of the first pioneers to invade a burn zone. Dozens of other wildflowers scent the air, their names a mystery to me. The fire has cleared the space for all these young plants to flourish.

Like an avalanche, the fire also opens the canopy and allows more sunlight to filter onto the forest floor. But fire goes a step farther than an avalanche: A fire recycles nutrients back into the soil. Trees like western larch can quickly colonize the ashy, mineral-rich ground. Larch is another species making a comeback in the Robert Fire zone, but is nearly absent here at Trapper, which burned too hot. The slow recolonization of the Trapper Plateau is one reason that, while fire on a modest scale can be beneficial, on a large canvas it can be disastrous. Fierce fires strip all vegetation from the land. Without trees, the soil is exposed. Already, erosion is cutting into the banks of Mineral Creek. This story is repeated across the fire-prone Rockies. In recent years, whole forests have been leveled. Severe fires can do more ecological damage than clear-cutting a woodland. Their watersheds deteriorate. Streams choke with silt. Fish are displaced. Dominoes fall.

After eight years at Trapper, however, some shrubs have taken root. Along the trail we feast on thimbleberry and raspberry, two closely matched fruits. The thimbleberries in the palm of my hand resemble raspberries, but have a richer, multiple-berry flavor—sort of like eating a mixed fruit gelato. Farther along, we come upon a stand of huckleberry bushes, which are rejuvenated by fire and are a favorite snack for moose and grizzly bears. I'm partial to them, too. I've had huckleberry pancakes, huckleberry jam, huckleberry pie (not just a piece of pie but the whole pie), and huckleberry chipotle sauce. If they made a huckleberry toothpaste, I'd be the first to try it. The flavor rests somewhere between blackberry and blueberry, and a strenuous hike can quickly become worthwhile when you come upon a patch.

From overhead comes the *rat-tat-tat* of a three-toed woodpecker tapping on a dead tree. Insects invade burned forests; woodpeckers and other birds follow. A new community is at work, and the forest will eventually revive. For decades, it will likely be a mosaic of species rather than a monotypic spruce stand. I'll have to return to see.

In moderation, natural disturbances add immensely to the fabric of the mountain ecosystem. Sherry turns to me and says, "If it wasn't for fires and avalanches, Glacier National Park wouldn't be as scenic." We hike out the last two miles to the van.

Nobody doubts that drought has helped spark the recent outbreak of fires across the western United States. Eight out of the past twelve years have been the hottest and driest on record. Yet the cause of the heat is hotly debated. Some claim the fire years have been fueled by natural climate cycles, like El Niño and the Pacific Decadal Oscillation. Others claim it is a clear example of global warming. We live in a polarized world: Typically the two camps take two opposing, mutually exclusive positions, turning every argument into a black-and-white exercise. However, in the case of forest fires, as in the case of melting glaciers, it is not an either-or proposition. Elements of both camps have merit. Background climate cycles, like the PDO, are acknowledged by most scientists, including Fagre and Divoky, as influencing glacier mass balance and fire frequency and force. Some suggest rising temperatures from climate change may become superimposed on these basic cycles, either amplifying or dampening the effects. With the general trend toward higher temperatures, the warm phase of the PDO may become warmer, and the cool phase may be less cool. Climate cycles are perhaps about to shift an octave.

Most scientists agree that modern climate change, as well as traditional climate cycles, is influencing weather. But there is a secondary debate about whether drought or poor management practices are primarily to blame for fires.

In the summer of 1910 much of the West was ablaze. It was the worst series of forest fires in a generation: A total of three million acres burned, leveling four towns and killing at least eighty-five people. Over 120,000 acres burned in Glacier National Park alone. After the horror of that summer, forest managers initiated a policy of extinguishing every fire across the West, thinking they'd curtail any chances of a runaway inferno like what they had just seen. One hundred years ago, knowledge of forest ecology was rudimentary. The new firefighting practice, named "fire suppression," was short-sighted. In the long run, the managed forests would become more susceptible to catastrophic fire than ever before. Forest stands that previously supported fifty to one hundred trees per acre, because they had been culled by fire, now had hundreds or thousands of trees crowded into the same area. And the forest floor was littered with snags and windfall, fuel ready to explode. By the tail end of the twentieth century, wildfires were sprouting up more regularly. They were more violent than before. In 2000, nationwide over seven million acres burned, more than double the western figure from ninety years before. The losses were climbing. Then, there was an effort to change policy in the 1970s and 1980s: a shift from fire control to active fire management. The most significant new measures were brush-clearing and prescribed fire. But for much of the Rocky Mountains, it was too little too late.

So those decades of fire suppression—and the build-up of fuel—are strong links in the chain of today's runaway fires. But how do these factors compare in rank to the forces of climate change?

"I'm not really against global warming," says Divoky, "but I don't see how you can distinguish between increased fire from global warming versus increases from less fire suppression on top of all that fuel."

A team from the Scripps Institution in La Jolla, California, has tried. Led by Anthony Westerling, the four-man team undertook a systematic analysis of forest fire activity in the West. The acceleration of major fires, beginning forty years ago, coincided with an increase in spring temperatures, which melted the snowpack earlier, followed by hotter summers. This new evidence points to climate change, not fire suppression or fuel accumulation, as the primary driver of fire prevalence and power.

To pinpoint climate's role, the Scripps team compared western U.S. fire history to spring and summer temperatures for the thirty-four years from 1970 to 2003. The increase in fire numbers and size were strongly linked to higher temperatures. The link was most pronounced at mid-elevation forests in the northern Rocky Mountains, the domain of Glacier National Park, where suppression has relatively little effect on fire risk. The wildfire seasons started earlier (once the snowpack melted); fires dragged on longer; and the season ended later in the fall. Overall, the duration of the wildfire season expanded by two-and-a-half months.

That rise in temperature, beyond the normal oscillation of the Pacific surface temperatures, spells global warming. Heat hovers like a warm blanket, drying out the wood. With the timber primed, all that is needed is a spark (lightning) and bellows (wind) to spread the flames. In Glacier, over one million acres is vulnerable. Lightning is inevitable, igniting ten thousand fires across the country each year. It has struck for millennia throughout the Rockies. The difference today is that, thanks to heat and fuel, ignition is more frequent. The burns are bigger and more common. Firefighting for U.S. wildfires now exceeds one billion dollars annually. The budget grows yearly. Part of that money can be added to the cost of climate change. It comes from our wallets. Not acting to stem carbon emissions has a huge price.

Still, not every scientist in Glacier is fully convinced. Dennis Divoky is cautious. He would like to see more evidence before conceding the debate on which element of the three—reduced fuel suppression, PDO, or climate change—is most to blame for the fires.

"If we started to see droughts running for decades—longer than the Dust Bowl years—then we might point to global warming," he says. Of course, it's possible all three elements may be at work at once.

The Dust Bowl on the American prairie was a time of severe drought, during a positive, warm phase of the PDO, running from 1930 to 1936 (in some regions up to 1940). The current drought of the American West could soon exceed that duration. In some eyes, it already has.

In the meantime, timber burns, releasing carbon dioxide into the air. Carbon storage by the trees becomes a carbon source for the atmosphere. The release of greenhouse gases from fires could further warm the planet and, in turn, that warming could prompt a climate shift that favors even more burning. It appears the future may end in fire rather than ice. Mark more frequent and extreme fires (and fewer glaciers and avalanches) on your calendar ahead.

Things Fall Apart

At Logan Pass, a broad saddle with parking for hikers and motorists at 6,648 feet (2,026 meters) on the Continental Divide, I check my stride, hesitating. I've just started the trail but must stop. I have forgotten my binoculars, I realize, and reverse course to speed walk back to the car. Sherry is already past the trailhead and keeps hiking, leaving me to catch up after my errand, an essential one because the Highline Trail, which follows the Garden Wall for seven miles to its destination at Granite Park Chalet, is known for wildlife sightings. That's our mission—spotting big game and checking for ice on Upper Grinnell Lake. Dan usually carries binoculars, too, but he's sequestered in the lab today, waiting for an ice-free window to access Grinnell. So I'm short two pairs. At the car trunk, with limited room in my small pack, I abandon a fleece jacket in favor of the field glasses. It's a brisk September morning, but I'd rather be cold than blind.

Back on the trail, I round the first corner and instantly run into a lady in white. Not a mountain this time but wild game the color of snow. A large female mountain goat—a nanny—guards the path, her single kid at her side. Against a backdrop of black cliffs that flank the uphill edge of the trail, the pair resemble ghosts—or, at the very least, trick-or-treaters covered with white sheets—tiptoeing out of the night. She has a classic goatee, like a billy, but the kid has a rounded chin. Everything is ivory, except their black hooves and horns. The devil in white.

In winter, herds of mountain goats huddle on windswept ridges, blending into the snow. This camouflage is their best protection against cougars and wolves, but they have another evasive skill: a near-acrobatic ability at climbing and descending precipitous terrain, even vertical cliffs. They go where predators fear to tread. They are gymnasts without the benefit of opposable thumbs. But they have another adaptation. Their hooves have a corrugated, rubbery sole, adept at gripping rock, and a sharp outer edge, useful for kicking any predators who brave a precipice while in pursuit. Despite their aptitude for ascents, inevitable and fatal falls claim many lives each year. Winter starvation, because food is hidden by snow, is another cause of death. In extreme cold, they mainly live off stored fat and the occasional huckleberry twig. Right now, at the end of summer, mountain goats have plenty of lichens, grasses, and herbaceous leaves and buds from which to choose. And yet, this nanny seems preoccupied with a likely patch of urine on the hillside. She licks it for the salt. When I approach, as close as forty feet, the nanny crosses to the downhill side of the trail, the kid above her, protecting it from a fall.

They disappear down the slope. It would seem my dash for the binoculars was unnecessary. The encounter had been up close, literally at my feet. I catch up with Sherry, who tells me mountain goats lose their fear of humans after a few generations in a national park like Glacier. Sometimes dozens are spotted at once. In summer, she says, they congregate at mineral springs to lick the salt. Walton Goat Lick, a salt deposit at a lower elevation on the east side of Glacier National Park, is a mecca.

But I'm happy to see mountain goats on Logan Pass, near the snowfields they call home. They are rarely far from a patch of snow. Even in summer, they favor water crystals, preferring to eat snow than to drink water. Like a few other mammals that are linked to snow and cold—lynx, wolverines, snowshoe hares—mountain goats are endemic to the alpine environment and may suffer most when the ice vanishes and snows continue to wane.

I feel a little guilty watching Lisa and Lindsey pass us on the trail with their enormous packs, standard issue for USGS personnel. They have the mother ships, and I just have a pod. Lisa carries photography equipment to do a special shoot farther along the trail. Lindsey plans to service an alpine weather station three miles ahead. They're in a hurry.

"No car batteries to lug today," shouts Lindsey, rounding a corner. "I have nothing to prove."

The trail ahead wanders in and out of the trees, fragrant with spruce sap. Mountains loom in the distance. Over my shoulder, Jackson Glacier, now seventh biggest in the Crown, glistens like a jewel. Following a contour, the north-wending path is fairly flat but crosses a steep slope. To the left, downhill, is dark forest; to the right, uphill, is green meadow. Here and there, patches of trees, often stunted, invade the alpine zone. At close to 6,800 feet (2,072 meters), I am near the limit of tree line. It is an inhospitable environment. Likely kept in check by both snow and cold, the conifers of the tree line cling desperately to life.

That Montana tree line is about average for North America.

Farther north, in Alaska, where the fingertips of deep cold reach lower elevations, tree line might be at a couple of thousand feet, sometimes at sea level. South of us, in Colorado, tree line is at 9,000 to 10,500 feet (2,743 to 3,200 meters) or so—approximately 3,000 feet higher than here. The rule of thumb is that for every 300 to 500 miles of latitude northward the mountain zones—alpine, tree line, and montane—begin a thousand feet lower in elevation. Glaciers also follow this rule, though loosely: On Mt. Popocatepetl in Mexico, glaciers begin at 14,400 feet (4,400 meters), while in Montana, the first glaciers appear at 7,000 feet (2,133 meters). In Glacier Bay, Alaska, glaciers appear at sea level. Remember: It's colder at either higher altitude or latitude. Mt. Everest, though only as far north as Tampa, Florida, might as well be on the North Pole.

I follow the edge of a stand of spruce as it dips into a ravine and races upslope, searching for water. Here, the sedges and grasses of the alpine meadow (also known as "alpine tundra") die back under the shade of the trees. This interface between conifers and alpine meadow species—called an "ecotone" by biologists—is like a tug-of-war from gully to gully, ridge to ridge. Competition for space is intense and dynamic. As if straddling a fencerow, trees and grasses make poor neighbors on an alpine slope.

I step out of the shade, and the trail erupts in the color of wildflowers—a palette at my feet, ranging from ivory to saffron to violet. The first flame is a rosy paintbrush, whose flower is actually scarlet, refuting its name. Hugging the path are pink asters, yellow arnica, and lavender daisies. A gorgeous display, considering we're just past the peak flowering time now in late August. The flower heads of false hellebore have lost their inflorescence but their huge pleated leaves crowd the trail. Underneath the hellebore, the heart-shaped leaves of an arnica plant catch my attention, and I thumb through my wildflower guide. Sure enough, the oil from the leaves of this yellow flower can be used to soothe sore muscles. I store a few in my daypack in case this twelve-mile trail gets the best of me. (Only later do I learn that the NPS discourages the picking of native plants.)

In all, I count more than a dozen species of wildflowers adapted to this land between worlds—the subalpine interface at the edge of tree line. Like the intertidal beach, another ecotone, the subalpine zone is only inhabited by a specially adapted crew. Most plants are perennials because the growing season is short; few annuals are able to complete their life cycles during one brief summer season. The dormant plants must endure a bleak and frigid winter on the order of an arctic December.

An odd wind, sounding like an old squeezebox, rustles up from below and climbs the valley, stirring the garden. In the distance, the mountains across the McDonald Creek Valley—Heavens Peak, Anaconda Peak, and others—look stern, immovable, and remote, but their flanks can funnel a canyon breeze. And what a scent it brings now: the thick fragrance of wildflowers from a dense meadow up ahead. To my mind, there is no memory more easily stirred, and more powerful, than a remembrance stimulated by the sense of smell. I am taken right back forty years ago to a summer clearing in the Adirondacks with eastern columbine in bloom.

Yet these are golden columbine, a similar scent but yellow instead of red. If I close my eyes, I can still see that scarlet field, patronized by bees and hummingbirds. Why do flowers play so deeply on the subconscious? Beauty perhaps. We crave beauty, then go out of our way to destroy it. But here it seems relatively untouched. Glacier, at the apex of three ocean watersheds, is still a beacon of wilderness, though its radiance is flickering these days.

Most of the subalpine flowers are yellow or white: arnica, spirea, yarrow. I wonder why this is so. Maybe insect pollinators are attracted to scent first and color second. Perhaps they are a little like bears in that way.

At the other end of the columbine meadow, nestled at the foot of an Engelmann spruce, I find one of the most beautiful blossoms of the day: a Mariposa lily. Mariposa is Spanish for "butterfly." The bloom is exquisite: A cup-shaped, yellow-and-white flower appears at the tip of a single, long stem. Native Americans and settlers ate the bulbs of this lily, which reportedly have a sweet, potato-like taste. As if living up to its name, the flower successfully attracts a small white fritillary, which—after one tentative flyby—alights on the silky hairs of the petals of the flower. The barter between nectar and pollen begins. I sit down to watch. For a lifetime, I've been partial to butterflies.

My first connection to nature was, at the age of six, collecting butterflies and other specimens for my "bug museum." This insect gallery occupied my parents' two-car garage in Baltimore, where dozens of glass jars and aquaria held treasures from my collecting expeditions to woods, ponds, and neighbors' lawns. Admission for the public was fifty cents—to cover the cost of lettuce and other foods—but neighbors who shrewdly alerted me to a praying mantis in their gardens got a free pass. Through trial and error, I learned the ways and wiles, the habits of many creatures—for example, my discovery that the predatory wasp, commonly called the cicada killer, returns from a kill with its prey and carries it into an underground burrow. I would wait behind a tree, then slam a jar over the hole, catching them both. I only kept each critter for a couple of days, then released them, except for butterfly cocoons and turtles. I am also partial to turtles and anything that hibernates. I was always elated to rediscover old friends in the spring.

Montana and Maryland share a few species that emerge after the snows. In both states, woolly bear moths freeze solid in winter and miraculously thaw out in March to continue their wiggle across the ground. Some say the thickness of their "fur" in autumn signals the severity of the winter to come. I've heard that woolly bears in Glacier are scantily clad of late.

The fritillary lifts off the lily and flits here and there—imperfectly making her way to the open meadow—to evade predators and me. She will mate and lay eggs and die. In spring, her offspring will evolve from larvae to green caterpillars to cocoons. From each chrysalis (pupa) will emerge a butterfly—a pinnacle of nature's design. I stand up, imbued with wonder. Nature is accessible wherever you look.

I traverse the finger of spruce, an island of trees amidst the sea of wildflowers. Under the canopy, I rejoin Lisa and Lindsey, who are finishing their midmorning tea break. The trail climbs a little so that in a hundred yards we are above the trees and crossing a true alpine meadow. We are up to our waists in cow parsnips, whose three-lobed leaves resemble a maple. The white flower clusters are shaped like little umbrellas. In the spring, grizzly bears feast on the succulent stems, which are also eaten by people of the Blackfeet Nation. I save my appetite for the thimbleberries growing in hummocks within the meadow. I pick two at a time: one for the larder and one for me. We all try to save some but they don't seem to stay in our pockets.

The trail ascends again, then cuts north, paralleling the knife

blade of the Garden Wall. This is an arête, formed when two glaciers ground the two rock faces, east and west, front and back,
like the pressure-flaking of an obsidian tool. The knife edge is narrow and runs sharply from Logan Pass to Swiftcurrent Pass; it is
part of the Continental Divide. Here on the west side, we cross a
ravine of running water. Snowpack, the last to linger on the slope,
feeds it from above. The vegetation is more lush along the stream
bank: sedges and potato-like spring beauty and mountain chives.
With a little imagination, you could lay out a banquet. Lisa and
Lindsey have already stopped to graze. We taste our share.

"This meadow, all of this wilderness, is rich," says Sherry, wav-
ing to a small herd of goats. She is slow to share her harvest with
man or beast.

"This landscape has been just as fertile for ten thousand years,"
says Lindsey.

"After the glaciers disappear, it won't be the same but we'll still
have wilderness."

Lindsay agrees: "Glacier National Park will be as natural as
it gets."

Just past the stream crossing, we come upon a field of "beargrass." Another misnomer: It is not a grass but a lily. Atop long
stalks nearly four feet tall, the white flower clusters are shedding
their star-shaped flowers. Hundreds of plumes are in their last
glory. The dead stems will persist through winter. Perhaps foraging on the remaining blossoms, two bighorn sheep stare at us
from the far side of the field. Two young rams with modest horns.
In winter, mountain goats forage on the tough evergreen leaves,
which form grasslike tussocks at the base of each plant. Bears use
the leaves as bedding material in their dens. Like the monument
plant, beargrass may go years without a bloom, then the commu-

nity, connected by rhizomes underground, puts forth flowering stalks at once. I happen upon a banner year.

Beyond the beargrass, several more fingers of spruce and fir reach uphill, each digit claiming a little more ground in altitude. None are broader than a few trees but they run up to fifty yards in length, and are known as "ribbons" of forest. At the tip of the ribbons, the living trees are often prostrate, hugging the ground, as a protection against snow and wind, and are distorted—what the Germans call "krummholz" (crooked wood). The boughs of the subalpine fir are oriented uphill, opposite the direction of the prevailing wind, like upended umbrellas. They are said to be "flagged" in this way, going with the flow to prevent being snapped in two.

The windswept slopes of the Garden Wall are harsh in winter. Winds in excess of 120 miles per hour (200 kilometers per hour) have been recorded nearby. Life here has its limits. Historically, deep and persistent snowpack has probably been a constraint on tree colonization in the subalpine zone. Now, with lighter snowpack, tree islands may reach higher each decade. With climate change, there may be less snow to keep the advance of timberline in check. Temperature will also play a role. And geology. Factors and interactions limiting or advancing the tree line will likely be complex.

Even below the krummholz zone before me, the ribbons are swelling with new trees. At the edges of the meadow, seedlings and saplings are shading out the wildflowers and grasses. The balance has shifted. Little trees dot the open meadow like the stubble on a young man's beard. Eventually, the ribbons will merge, the stubble will grow, and mostly forest will hug the hillside.

The tug-of-war between trees and tundra is happening all over the Crown. I spent a day recently with Dan Fagre and Lisa

McKeon on one of the battlegrounds, "Preston Park," below Pie-gan Glacier, on the east side of Montana's Continental Divide. (Within Glacier National Park, large alpine meadows—or fields of alpine tundra—are often called "parks.") On this park within a Park, Fagre and McKeon surveyed trees growing in ribbons at the edge of three beautiful meadows. They were examining individual trees to assess damage from the winter snowpack. Many were topped off or buckled or bent; some will not survive. Fagre announced that their survey hopes to pinpoint the spectrum of factors that limit the advance of trees. He's uncertain of their ranking in importance at any given site but knows that soil, debris flow, wind, avalanches, and topography play a role along with climate. Lisa's and his tree study will perhaps nail down other suspects, including snow loading and, behind that, the Pacific Decadal Oscillation. Visiting the study site on the opposite side of the Divide demonstrated to me how widespread the problem is—the problem of tree migration. By some estimates, Glacier has already lost around four percent of its alpine habitat to encroaching forest.

Krummholz grows into odd shapes and, as a pioneer, doesn't always survive at the leading edge of uphill colonization. Ahead of me now, bleached stumps are strewn on the ground, victims of false starts, while new trees keep charging ahead like infantry on a battlefield. In the middle of one tree clump is a lone dead trunk, forked and crippled, towering above the prone subalpine fir. Surrounded by living shrubs, the ghostlike center looks like a timber atoll. This is the corpse of a whitebark pine, most likely killed by blister rust, an exotic parasite that has devastated whitebark communities throughout the Rocky Mountains, leaving sticklike forests behind. The loss of whitebark pine has a trickle-down effect through the food chain. Grizzly bears favor whitebark pine nuts, gorging on them in prodigious quantities. They often steal pine seeds from ground caches hidden by Clark's nutcrackers or from middens hoarded by squirrels. The connections are intricate. Each player in the alpine world is dependent on a pyramid of workers and plants below. Thus, any disruption—like the march of trees uphill into alpine meadows—is a challenge for all creatures, great and small.

The pressure on neighboring communities has become intense in the face of an advancing tree line. While the first impact is the crowding out of alpine meadows, this loss is not simply an end to a few wildflowers it comprises a potential loss of myriad species as well as ecotones, those habitat edges of such importance to wildlife. Timberline is not a straight line, but a jagged and wavy delineation. So there is a loss of considerable interface area. With the reduction of meadows and their ecotones, researchers anticipate the exclusion of many foragers—bighorn sheep, mountain goats, mule deer, and elk—that depend on tundra for grazing. In the shrinking world of the alpine meadow, plants will certainly be the first to suffer losses, closely followed by these herbivores. Rare species will be the most vulnerable. Genetic diversity will lessen as the tundra is replaced with trees since montane forests have fewer species than alpine tundra.

The Mountain Geodynamics Research Group, led by David Butler, George Malanson, and Steve Walsh, and guided by Dan Fagre, has been tracking the changing tree line in GNP for over twenty years. Inspired by the work of R. A. Rockefort, who, in the early 1990s, documented the movement of trees to higher elevations throughout the West, the team explored the "possible sensitivity" of trees to the Montana climate as it warms. While the invasion of tundra by trees was an expected outcome, the team took nothing for granted. They measured and tallied all possible influences of advancing trees, ranging from soil type and landslides to temperature and snowpack conditions. To catalogue current timberline, they employed remote sensing and photographic techniques perfected by Fagre on his glaciers.

Beginning with photographs from around 1910, they repeated the shots—frame by frame—and compared the tree lines. In some locations, the trees had advanced upslope by thirty, sixty, ninety, or more feet. In other cases, the tree cover had become more dense; however, in many photographs there was no change at all. To the scientists, these mixed results suggested that general predictions were not yet feasible. Too many factors were at play.

In particular, they suspected rock slides and uneven terrain, as well as universal tree conditions like soil and seepage, could create "tree lines" that are not shaped by climate. Even trampling and wallowing by sheep and goats can affect tree seed establishment. With all these influences at work, they were not surprised to find ribbons of forest manifesting well below the snowpack. Tree lines were erratic at best.

Thanks to such wild cards, the Montana team concluded that any given stretch of tree line was not a dependable "indicator" of climate change. Unlike glaciers, which they consider to be highly responsive to climate warming and thus good indicators, local tree lines are too unwieldy. In other words, the appearance of a few seedlings or saplings uphill from the old timberline in just one spot is not conclusive, but a broad uphill march—say, a mile-long swath of advancing trees—could be attributable to climate change. A shrinking snowpack (and elevated temperatures) will likely prompt trees to climb.

David Butler and his colleagues concluded in their report,

"Tree line in the Park viewed by our grandchildren will not be the tree line we have studied over these many years. The world is changing because of climate [warming], and tree line is a part of that world of change."

In the meantime, the mapping of timberline undertaken by the team will serve as a baseline for further perturbations, either advances upslope or any retreat. The tree lines of our grandchildren's day will furnish the proof.

At this point on the hike, Lisa and Lindsey rush ahead. Sherry stays close for the trek toward Granite Park Chalet and any tangents in between. She is in her usual good humor.

"I've got your goat," she says, slipping by me on the trail and taking the lead.

I couldn't protest. Straight ahead, a billy goat stares us down.

She waves to him. "They don't like men," she says with a grin, joking.

By now, we are detouring east off the Highline Trail past a small promontory of rock and tundra opposite Haystack Butte. This side path, actually a game run frequented by Sherry's goats, takes us just above tree line below the Garden Wall. (On the way, we can see the top of Gem Glacier peaking over the Garden Wall.) To the west, the Haystack summit levels off at 7,600 feet (2,316 meters), approximately 1,000 feet (304 meters) above us. After a series of gentle switchbacks, climbing 300 feet (90 meters) or so, we reach our destination: a compact weather station, its wind vane wavering in an open meadow. Like the Sperry station, the weather gauge is an aluminum box, about the size of a small suitcase, but is outfitted with antennae for transmitting data back to base. Lindsey opens the padlock to reveal the varied instruments inside, ranging from temperature, wind, and humidity sensors to a barometric pressure gauge. Next, she ascertains that the sensors and solar panels are working properly. She gives a thumbs up to Lisa. Together with the five other USGS weather stations in the Park, the small box has helped to pinpoint the level of climate warming in western Montana.

Later, Lindsey glances upward to the crown of Haystack Butte. "We're also monitoring four summits," she says, "for temperature, yes, but mostly to document their vegetation—to see if plant diversity is changing as the climate warms."

Those four lofty summits, from Dancing Lady to Seward Mountain, are part of an international network of mountain peaks being studied simultaneously on five continents. The surveys follow the protocol of the GLORIA program—the Global Observation Research Initiative in Alpine Environments, headquartered in Vienna, Austria. The plan calls for a multi-summit approach, typically four mountains in each target region, which are canvassed every five years to see if there is a change in soil temperature and species diversity. So far, over ninety-four regions have been established, ranging from the French Alps to the Urals (Russia), from Kanchenjunga (Nepal) to Mt. Kenya, from the Brooks Range (Alaska) to Huascaran (Peru).

The Austrians anticipate an *increase* in the number of plant species on summits as low-altitude vegetation moves upward, under the pressure of warming, and overtakes the relatively species-poor, high-altitude zones. There was already evidence for their hypothesis when they began GLORIA in 1998—in the Rhaetian Alps of Switzerland. Most of the first work was done in Europe. Many alpine plants in the Mediterranean region (e.g., the Sierra Nevada of Spain) and elsewhere were discovered to be rare and endemic, so displacing or destroying them has consequences for genetic diversity. This is a crisis worldwide, particularly for alpine medicinal plants, which in Tibet, for example, comprise seventy-six percent of traditional remedies. Cures for untold diseases in western medicine may be found on mountaintops. Besides offering such "ecological services," the fragile alpine zone comprises one of the last untrammeled ecosystems on earth—reason enough for preservation.

As it happens, summits are ideal laboratories for studying climate change. Peaks are exposed to all aspects of the compass and, being equally exposed, each quadrant has similar conditions, other than sunlight. North slopes get less sun. But, for the most part, climatic conditions are largely defined by altitude. The plants themselves are long-lived, slow-growing perennials and are limited in their respective altitude tolerances by temperature (and by a few other factors). Therefore, with warming, it is expected that plants would migrate upward to cooler positions. As such, the alpine zone and the nival boundary (the band just below the permanent snowpack, which is the upper limit of plants) are good indicators of climate warming. Much more so than tree line. The zones occur on most high mountains. Summits are globally distributed, another advantage for a series of outdoor laboratories. Also, mountaintops are not prone to severe disturbances, such as avalanches, or to much traffic from the public, making them ideal observation posts. The climate signals can be read without interference.

Dan Fagre and Lindsey Bengtson have adopted the GLORIA protocol and put it to the test in Glacier. They instituted their first survey in 2003–04 by cataloguing all summit species, and repeated the daunting task in July 2009. The four peaks chosen represented the elevation span from grass meadows to snowpack at the highest summit. With a team of botanists, they identified every single species and classified their abundance from rare to dominant. At least four wildflowers were common on all the mountains—alpine smelowskia, sky pilot, cut-leaved daisy, and shrubby cinquefoil—but each peak had its own profile of plants. The highest summit (Seward Mountain) had the fewest species— only thirty-nine. (The greatest diversity was found at the two mid-level peaks, averaging seventy.) What was most important, however, was how the diversity of each location changed over time.

The results were stunning. While Dancing Lady, the lowest summit, was fairly stable with only an eight percent increase, Seward Mountain (the highest) had fifty-one percent more species in 2009 than in 2003–04. The mid-altitude peaks were colonized by twenty percent and twenty-five percent more plants, respectively. The explanation seemed obvious. Some observers were quick to claim plants had to be ascending the mountains because of global warming.

Not so fast, warns Lindsey. She is suspending judgment until the next full survey in 2014. "We saw a lot of species turnover in 2009," she says, "but we need another season of research to be sure of the reason." She cautions that other processes besides a shift in climate may prompt uphill colonization: for example, unusual weather variability from year to year. During 2003, the year of major fires, including the Trapper blaze, the summer was hot and dry. "Perhaps some plants died from the heat before we identified them," says Lindsey, "and other species took their place in the intervening years. Meanwhile, 2009 was unseasonably wet, so certain vegetation may have thrived. This variability isn't climate; it's simply the weather. We have to look for longer trends."

Summers were also warm and dry during the intervening years, a heat wave shared by Europe, where another GLORIA site was being resurveyed. Brigitta Erschbamer of the University of Innsbruck and her Austrian team had documented four summits in the Dolomites of Italy in 2001; they returned in 2006. Like Glacier, their tallies were in line with the GLORIA hypothesis. Species richness climbed in the upper alpine zone and subnival zone, just below the snow, by ten percent and nine percent, respectively. Only modest increases were found farther down, including at tree line. The greatest gain upward happened at the higher summits, where alpine plants are most specialized and sensitive. Other disparities happened from peak to peak. At the lower summits, the newcomers were identified as invaders from tree line and the lower elevational bands. At the highest summit, the transgressors hailed from alpine tundra below. All mountain zones gained species; there's a reservoir of plants below that are ready to recolonize upward to remain in a comfortable temperature zone.

Latitude mimics altitude. Like rising temperatures at high elevations, it is getting warmer farther and farther north. Alberta is looking more like Montana. Some arctic and alpine plants once found in Glacier are now confined to mountains in Canada. Others are declining sharply. Specifically, plants like northern gentian *(Gentiana glauca)*, which lives at the southern extent of its range in Montana, are dying out—for gentian a decline of nearly forty-four percent in little over a decade. The wildflower requires a cool climate.

So, why is greater diversity on a mountaintop a problem? Simply put, the original, uppermost species may be displaced. For the time being, however, it appears many alpine species are staying put amidst a plethora of new neighbors. But, eventually, some may forfeit the space. Perhaps the 2014 GLORIA survey will confirm the winners and losers. In a perfect world, the plants could keep climbing—colonizing higher—as the mountain climate warms. But they are already at the altitude limit: Above the peak is thin air.

Thanks to the geometry of summits, habitat shrinks as elevation climbs. Alpine species are thus susceptible to an elevational squeeze, whereby escape routes to cooler habitats uphill become dead ends. Throughout the world, plants and animals are fleeing the heat. But, as tropical tree frogs and toads have discovered in Guatemala and Costa Rica, by migrating uphill they eventually run out of room. Central American peaks are no longer cool. Even Mediterranean mountaintops are hot. It has been predicted that low-altitude mountain ranges, such as the northern Apennine in Italy, might lose fifty percent of their species by 2100. The extreme European summers make this ever more likely. The first decade of this century was the warmest in the European Alps since 1500.

We descend from the weather station and gain the Highline Trail for our hike north to the Overlook, the photography and observation stance Lisa has in mind for the day. Lindsey and she race ahead, as we dawdle, putting an hour between us. We are at the height of the diorite sill, a black igneous band that permeates the Park. The path wanders back and forth through the tree line, each band of conifers pungent with sap, each open meadow loaded with wildflowers. It's an even grade, with little uphill, but around a bend in the diorite sill we dip downhill toward a ravine filled with loose boulders, up to the size of a wheelbarrow.

From below, a high-pitched squeal stops me in my tracks.

"Eeeenk!"

Five seconds, then another piercing *"eeeenk"* echoes off the sill. The nasal call announces a community of pikas, the small, high-altitude relative of the rabbit, which occupies talus slopes, or rock piles, throughout the mountains of the American West. The talus terrain above tree line becomes snowpacked in winter and inhospitable to most creatures. Yet it is essential to the pikas' survival. They can hide beneath the snow and rocks. Either a territorial display or an alarm, the ear-piercing whistle likely cautions dozens of other little pikas living in the jumble of stone.

Sherry spots the siren and points. On a modest boulder in the middle of the 120-foot-wide rock pile sits a thick-furred mammal, resembling an oversize hamster, about six inches long. This is the pika and its village: a cold-loving species that chooses the rock pile's cracks and crevices for its year-round, defensive home. In addition to providing warmth from the cold, the crevices are its only partial protection from predatory weasels. Rock piles also furnish a cooling-off place in summer. Because of its thick fur and diminutive ears, it is prone to overheating. In cages, pikas have died when exposed to temperatures as low as 78 degrees F (25.5 degrees C) for only a few hours. The pika dissipates little heat through its tiny, rounded ears compared to the related rabbits and hares. But like them, it has eight incisors; rodents only have four. The pika, like the mountain goat, is a quintessential alpine animal: It doesn't hibernate, staying active all winter long underneath the insulating snowpack. To survive the frigid winter, it maintains a high body temperature, which it can't turn off in summer.

The shrieks from the rock pile seem to have alerted two smaller pikas that now surface on the northern periphery, next to a meadow. They raise their heads as if to listen. Each carries a bundle of grasses in its teeth, crosswise like an ear of corn. The pair are ferrying the stems to caches above and below the broken rock to cure the food for winter. Little hay piles are scattered around the village. Every summer, each haymaker stores several bushels of greens. In winter, the pikas, each solitary in its crevice, will eat through their stores slowly, under the cover of rock and snow. Right now, the two haymakers dash beneath two boulders to make a stash, which they will defend, each alone. While solitary in repose, pikas aren't just self-interested, however. To distract a weasel, at a time like late August, several pikas may run wildly about the rock pile, perhaps a form of schooling behavior so that the weasel is overcome by choice. Such altruism, though risky, makes sense if the village is composed largely of kin. That seems likely with the next rock pile hundreds of yards away.

Another haymaker emerges and scampers off to the meadow. "Oh, man, that pika is a busy rabbit," says Sherry.

Not surprisingly, the pika is highly vulnerable to climate change. As the summers heat up more, they could overheat and die. Even more frequently, pikas could miss out on valuable foraging time when cooling off in their talus. As the snowpack shrinks, they could freeze in winter. While more mobile mammals are expected to migrate to more northern latitudes, pikas are homebound. They live in restricted, isolated rock patches, essentially alpine "islands." For the same reason, they cannot move upslope easily to fight the heat. They can't burrow, so they need a ready-made winter den. That means talus. In Glacier, talus is only an occasional punctuation point on the mountain page.

With no ticket out of their predicament, some populations are vanishing from the western United States. In the Great Basin Mountains of Nevada and Oregon, local extinction rates of pikas have climbed almost five-fold in the last ten years. Climate change is a likely suspect. Recently, habitat loss and degradation have generally become the most common cause of wildlife species decline and extinction worldwide. Bulldozers and chain saws come to mind. But here is habitat infringement likely brought on by temperature—because of carbon emissions, because we drive too many cars and don't demand cleaner factories and power plants.

In the Great Basin study, the rate at which pikas have migrated up mountain slopes has increased eleven-fold since 1999. In the past decade, the species has moved uphill at a rate of around 475 feet (145 meters), compared with around 43 feet (13 meters) per decade during the twentieth century. You can only move upstairs for so long until you're at the roof. And pikas can only move at all if there's rock rubble to be had. Be certain: Any migration by them should bring us concern. The climate-sensitive pika has become another early warning signal of planetary distress.

In 2009, the USFWS began a review of the pika's predicament to see if it should be protected under the Endangered Species Act due to rising temperatures. The petition pointed to its inability to survive even a small elevation in temperature. An initial finding one year later said that protecting the pika "may be warranted because . . . of effects related to global climate change." If listed, the pika would become the second animal outside Alaska to get protection due to climate warming. (The wolverine was proposed for listing in 2012 due to reduction of snowpack and other factors; in Alaska, the polar bear was listed as threatened in 2008 because its sea-ice habitat was melting away.) But, in February 2010, the USFWS ruled that the pika was not threatened by climate change after all because it could actually survive temperatures that the EPA projected would climb by 5.4 degrees F (3 degrees Celsius). The federal government also decided that the pika could survive the loss of snowpack, even though scientists had claimed the species depends on it for shelter and defense. The USFWS managers wrote, "The American pika has demonstrated flexibility in its behavior and physiology that can allow it to adapt to increasing temperature."

Even if the USFWS had listed the pika, it's not clear whether the government would be compelled to cut carbon emissions as part of the recovery plan for the species. When the polar bear was listed, the Bush Administration exempted greenhouse gases from control under the Endangered Species Act. The Obama Administration hasn't changed the policy. From the stonefly to the cutthroat to the pika, climate-sensitive species have been turned down for protection. Politically, their listing may be too hot to handle.

The sentinel on the rock pile emits one more shrill cry and vanishes into the labyrinth of rock, perhaps to avoid the heat of midday. We drink thirstily from our water bottles and Camelbacks, and move on. Yet the discussion on the trail is all about creatures facing a changing world—stoneflies, bull trout, pikas, mountain goats, even ptarmigans—the elusive mountain birds known for their varying camouflage. When danger approaches, ptarmigans freeze instead of flushing, relying on their cryptic coloration—mottled brown in summer, brilliant white in winter, which makes them nearly invisible against the snow. Changes in snowfall and other climate effects are making their alpine habitat unsuitable. Already, in Colorado, ptarmigan eggs are hatching twelve to fifteen days earlier in the spring. An earlier hatch is also likely in Montana. From invertebrates to mammals, warming temperatures and shrinking snowpack spell trouble for the entire menagerie of the Crown.

We cross an empty streambed, running dry from the early exhaustion of snowpack above. Then the trail dips down along a break into a ribbon of thick subalpine fir and pine. Sherry says subalpine and montane forest is the domain of the pika's cousin, the snowshoe hare. This rabbit is the primary herbivore of both types of boreal forest. Like the ptarmigan, the snowshoe hare is a "varying" species: It has a cryptic coat of brown in the summer, which it molts in favor of white fur in late autumn. Snowshoe hares are snow-loving creatures and have the big hind feet to propel them swiftly over the surface—to escape numerous predators, from great horned owls to mountain lions. The biological clock that prompts the change in coloration is triggered by the length of daylight. As days shorten in the cusp of late fall, the molt begins. This affinity for sunlight, rather than for a background of snow (like the background stimulus to a chameleon), may be the hare's undoing. In the face of climate change, with snows starting later in autumn and ending earlier in spring, the hare is exposed: a conspicuous white rabbit against a backdrop of brown or green. A hare that's the wrong color stands out like a target.

Enter the lynx. The life history of the lynx is tied inextricably to the snowshoe hare, which it hunts with almost single-minded focus. They occupy the same wooded habitat, most often lodgepole pine, with deep snow cover in winter. Together, they follow a closely matched population cycle that is most pronounced in Canada and dampened somewhat in Montana. Lynx numbers ascend and plummet in tandem with hare quantities, peaking just after the hares skyrocket, when the food supply is robust. Afterward, when hare numbers fall precipitously, lynx females do not become pregnant. The hares recover. It is a boom-and-bust cycle. Like most aspects of ecology, the deeper you delve, the more complexities you unearth.

Nearly the size of a bobcat, the lynx has longer legs and bigger feet, serving as its own brand of snowshoes. The only animal faster on snowpack is perhaps the mountain lion, which will kill a lynx to curb competition for the widely fêted snowshoe hare. However, deep snow will typically exclude the lynx's main competitors—coyotes and bobcats—from its high-country habitat. We aren't likely to see these wary predators today, but it's thrilling and comforting to know they still run wild in Glacier. One of their favorite prey in a brown coat, probably sits under a fir branch, within our peripheral vision but camouflaged and hidden from view.

Consider that summer coloration for a minute. Brown fur is the rabbit's summer armor. But if a white coat were worn in summer instead, the hare would be easy prey. Similarly, with mismatched fur in October and April, when snows are now often lacking, that white rabbit is vulnerable. "It's a bonanza for predators," says Sherry, "but we need hares. Not overhunting the defenseless ones. They furnish food for nearly every carnivore in the forest. Without the snowshoe hare, the ecosystem shatters."

Flying through the snow, the lynx is a singular cat, so specialized in its menu that it has a precarious place on the planet. The destinies of predator and prey are linked. If there's more rain and less snow, oversized feet won't do either of them a lot of good. Less snow might also spell more predation and competition for the lynx. To locate proper habitat, they will have to move higher or farther north. But corridors for migration often don't exist.

Since 2000, lynx have been listed as threatened, even though climate change was not considered. Numbers continue to drop. Less than a thousand lynx are estimated to live in the lower forty-eight states.

We are close to the cutoff to the Overlook Trail now, and we see a middle-aged woman (from Florida) and her teenage son come barreling down the path. Breathless, they tell us that when they were just perched on the Overlook, scanning Grinnell Glacier, a wolverine ran up the trail and scooted over the edge, body-surfing on the snowbank down to Upper Grinnell Lake. A descent of 1,600 feet (488 meters). A bold move from a tenacious beast, pound for pound the scrappiest predator of the mountains. (In Canada, a twenty-five-pound wolverine once brought down a three-hundred-pound caribou.) Like lynx, wolverines are dependent on heavy snow. Although their thick coat remains brown in winter, evidence suggests they are also struggling with climate change. The Florida teenager shares some of his knowledge with me; he had written a school report on wolverines in middle school. Wolverines evolved for life on the snowpack, he says, buoyed by outsized feet. But, as snow declines throughout most of western U.S. and Canada, wolverine numbers are falling. (Fewer den sites, available in the snow, reduce reproductive success.) In Canadian provinces where snowpack levels are dropping the fastest, wolverines are disappearing most rapidly. Perhaps no more than three hundred populate the lower forty-eight.

Wolverines defend and roam over impressive territories—males range over several hundred square miles. Nothing gets in their way. Take a radio-tagged male, known as "M3," for instance. During a recent January, he was headed toward British Columbia and the tallest mountain in Glacier, namely 10,466-foot Mt. Cleveland, stood between him and his objective. M3 did not hesitate. He scampered up the snow-clad south face, all 4,900 feet (1,490 meters) of it in ninety minutes. Then, he glissaded down the other side.

Reduced snowpack may threaten the wolverine on two more fronts: food and dispersal. They may find it more difficult to locate prey, especially carrion. Young wolverines may be limited in moving to new territory without deep snowpack for travel. Population numbers in Glacier are estimated at forty-five individuals. Remarkably, because of their large territories, they are seen quite often. We have just missed one of these by a nose, or rather, by tooth and claw.

At noon sharp, Sherry and I rendezvous at the trail junction. The climb to the Overlook cuts diagonally across scree and tundra, high on the Garden Wall, to a small gap in the Continental Divide, a chink in its armor. The path and the game trails that crisscross it are perfect habitat for mountain goats, but halfway up we have a different encounter. Three bighorn sheep, all young rams, stand on a rock outcrop thirty yards below us, munching on grasses, seemingly oblivious to our presence. In the open like this, the rams have a clear view of any approaching carnivores. But the encroaching forest will give greater cover to their predators—mountain lions and wolves—rendering them more vulnerable to attack. Less safety from predators will mean less game.

Yet, today, here at the beginning of the twenty-first century, the Crown possesses nearly all its magic. A visitor who is willing to step into the backcountry is guaranteed to see some of America's last big game. They are all here. Keep your eyes sharp. Even from a car, you can view mountain goats, deer, bighorn sheep, moose, elk, and bears—grizzly if you are lucky, which is often the case.

This is the second American Serengeti: Only in the Lamar Valley of Yellowstone can the diversity and numbers compare.

Glacier is the one place where I can live in the present, past, and future all at once. Here I am, wrapped in its majesty, and yet I'm at once envious of my grandfather's day and guilty over what my children will miss. We watch the three rams trot off down a game trail. After a bit, they turn a corner and vanish from view. I wonder if they'll make it through the winter. And the warming. Just now, we don't know the end of their story.

We traverse above the bighorn rocks, past an island of stunted trees, toward the cliffs that jut out below a pyramidal horn. From there, the faint trail cuts southeast to the saddle between a false summit and Mt. Gould. Under a bright blue sky, the USGS researchers are first to reach the saddle, a mere dip in the arête. The Overlook stance is just east, below the crest, but while Lindsey and Lisa quickly scamper down to the perch, Sherry and I pause for a moment to take in the view from the Continental Divide. Jagged mountains stretch in every direction, in rows like the tiers of teeth in the jaws of a shark. I don't know all their names and I christen the most serrated as "Snaggletooth." I'll never make it to the top. Another mountain to go unclimbed to keep the mystery fresh.

On a lark, I pour some of the water from my plastic bottle directly on the apex of the Divide. It is amusing to imagine half of the cup flowing east to the Gulf and half tumbling west to the Pacific. I'll have to try the trick on Triple Divide Peak sometime and claim the Arctic Ocean as well.

When I catch up with Lisa and Lindsey, they have already exchanged the memory card in the Extreme Ice Survey camera mounted below an overhang in the mountain wall. Beneath our feet is a breathtaking bird's-eye view: Looking down, I take in the panorama of Grinnell Glacier and its iceberg-strewn lake. They fill my vision. In the foreground is another wisp of ice; I realize that the tail of Salamander Glacier sits just below us to the right. From this distance, I cannot tell if either glacier has foreshortened in the three years since my first visit. But Grinnell looks small, as it claims only half of the broad rock cirque that it once filled. It has drawn back into itself like a bear in a den.

Yet the real news is the condition of the lake. The brisk wind shuffles small waves across the surface. The lake is free of surface ice. Even the icebergs float freely. They would present no impediment to the inflatable boats. Dan Fagre will be pleased.

"We've been waiting all summer for the lake to be free," Lisa tells me later.

"If it refreezes tonight, he'll never believe us," says Lindsey.

Just to make sure, Lisa clicks off several photographs of the glacier and lake below for her boss. They will constitute proof that the lake is clear. These images will also be compared to Overlook shots from previous years to track the movement of ice. The first such archival photograph was taken around 1940. It shows no icebergs in the newborn lake; the pool is considerably smaller. For over seventy years, the Overlook has provided such surveillance— a witness to the evolving climate.

After a half hour on top, I dash down the spur trail toward Granite Park Chalet. Lisa and Lindsey will stay behind in an attempt to fix the Extreme Ice Survey camera, whose mount has been pulverized by a falling rock—second year in a row. Sherry will return to Logan Pass. I race along, favoring a bad ankle somewhat and running a little recklessly. I'm giddy. The day has been flawless, the climb energizing, and it's all downhill from here.

Quickly, I regain the main Highline Trail and head north toward the chalet, which stands like a stone hub about a mile in the distance. Three trails converge on it like spokes in a wheel; I just have to get to the center. I don't like hiking alone in Glacier, and it is not recommended, but I anticipate soon joining the throng of hikers heading down to their cars in the late afternoon. A few minutes alone is not a problem, I think. Pepper spray hangs from my belt, after all. It has proven to be the safest deterrent against bears. A little comfort. I figure I am safe.

Crossing the first finger of tree line, I nearly step on a pile of bear scat. The sheer quantity makes it unmistakable. That and the huckleberry seeds. I clap my hands several times, loudly, to alert any predator of my presence. Then, I start singing, which would scare Ursus major out of its constellation.

Hiking the Highline Trail requires the suspension of fear. You imagine an empty trail ahead, even if you suspect it is not. At all costs, keep nerves at bay. The dense trees obscure my view.

I cross from the spit of tree line into an open, sloping meadow and am immediately relieved. Now I have a 360-degree lookout. Up ahead are four hikers—reinforcements—gawking at another patch of trees. I join them, but the ashen looks of their faces leave no room for a greeting from me.

The flash of brown on the steep slope below is quick and mesmerizing. The grizzly is so big it looks like two bears coming. I stop in my tracks.

The grizzly slows down slightly. Her gray-peppered hump flexes. Muscles ripple across her back and arms. She could be here in a second.

Suddenly, she pulls up short, about forty yards away. She stands her ground, swinging her head from side to side, sniffing. Then she looks over her shoulder, down the slope. Two mottled cubs emerge from the twisted trees behind her. A mother grizzly and her young—exactly the combo you want to avoid. Next, a third cub ambles into view. Triple threat. Luckily, we have come upon a field of glacier lilies. The bear family is more interested in digging up lily bulbs than countering a threat. She backs off and rounds up her offspring. They return to digging with their oversize claws. Her aborted charge happened so fast, I never had time to pull the pepper spray from its holster.

The five of us tiptoe to the other side of the meadow, actually an old avalanche path that has grown over with plants. Grizzlies and wolverines frequent avalanche tracks, not just for wild-flowers and berries but also for carcasses of animals that have been killed by snowslides the previous winter. In spring, grizzlies patrol the chutes like bloodhounds; they can smell carrion through ten feet of snow. Rangers call avalanche chutes "bear elevators" because of the speed a grizzly can climb a hill. On flat land, a motivated grizzly can outrun a thoroughbred horse for a limited distance (and easily a human). The bear can nearly match this gait on a rise. We were lucky she was bluffing today.

During the fattening-up season, a grizzly may consume up to 35,000 calories in a day, when flower bulbs and roots are high on the menu. In search of these delicacies, bears till swatches of meadow to bite off bulbs and taproots from underneath the sod. The process of turning over the soil increases the available nitrogen, nurturing next year's glacier lilies. Each spring, grizzlies return for even more, fatter bulbs, thanks to their gardening. Whitebark pine nuts and cutworm moths are also high-fat favorites.

By the time they retire for the winter, each grizzly adult may have gained more than a hundred pounds. The bears dig dens in hillsides, sometimes thirty feet deep, moving several tons of earth. Snow furnishes insulation, too. Grizzly bears are denning later in autumn and emerging earlier in spring. This is partly due to climate change but, at least in Yellowstone, grizzlies are consuming elk carrion (killed by reintroduced wolves), a menu that keeps them active on the cusps of winter. The consensus among scientists is that, in the short run, climate warming will not threaten the bear populations in the Rockies. Grizzlies are opportunistic and omnivorous, it is reasoned, and will adapt to alternative food sources. There is a concern, however, that abbreviated denning will result in greater bear-human conflict in the spring and fall when natural food sources are scarce. Grizzlies will likely descend to mountain valleys searching for a meal and will encounter people. Like all mountain species, grizzlies should continue to be monitored as the region warms, to confirm they are indeed safe from climate change. There are fewer than 1,200 grizzlies in the continental U.S.; twenty-five percent of those are in Glacier.

The north side of the meadow is adjacent to Granite Park Chalet. When we gain that safe footing, we turn around to scan over the bear family, from a healthy distance. The three cubs are frolicking on our trail now, having climbed the thirty-degree slope in a flash. Their mother is leisurely following them, seemingly oblivious to another coterie of hikers now coming down the pike. We wave our arms and shout. The two hikers see the cubs and run away—another mistake in any bear encounter. A running human resembles delectable prey.

By now, a crowd has gathered next to the chalet watching the bear family. One of the crowd is a park ranger, who cautions everyone to stay put. Personally, after the close call today, I'm willing to remain at the chalet, a stone fortress, without budging for a week.

Granite Park Chalet has had good luck with bears for at least forty-five years. But, the ranger tells me, in 1967, a rogue grizzly got the best of Julie Helgeson, nineteen, and her companion, Roy Ducat. They were camping close to the chalet, only about five hundred yards away. Both campers slept out in the open, without a tent. At one o'clock in the morning, a grizzly appeared out of nowhere, lumbered over their sleeping bags, slapped both campers, and bit deeply into Ducat's right shoulder. Showing restraint, Ducat tried to play dead. The biting stopped. He opened his eyes to discover the bear standing over his helpless girlfriend and slashing at her arms and face.

Ducat could hear bones shattering under the power of the bear's jaws. Helgeson kept screaming, "It hurts. Someone help us!"

At first, Ducat was too wounded, too in shock, to respond. Gathering his resolve, he ran uphill for help. Shortly afterward, the grizzly lumbered off. The chalet manager sent out scouts; however, it took an hour for the lodgers at the chalet to bring a makeshift stretcher for Julie. They had first tended to the young man, according to a triage system that first treats the patient who is most likely to survive. By the time she was treated, it was a miracle Julie Helgeson was alive. She had deep puncture wounds to her neck and chest that were bleeding severely.

By 4 A.M., Helgeson was dead. Her body and her boyfriend, alive but in critical condition, were evacuated by helicopter. Ducat would survive. But the story doesn't end there. On the same night, six miles away at Trout Lake, another nineteen-year-old named

Michelle Koons was also mauled, by a different bear. (Later, at least two bears were shot.) Michelle died of her injuries, too. Until that night, the Park had never witnessed a fatality from bears. Since then, there have been eight more deaths over forty-plus years. Some authorities believe it is because humans have encroached on the wilderness. It's all probability, they say: More bear encounters make lethal incidents more likely. While grizzlies have become a symbol of wildness, they still inspire fear and deference.

After hearing the story of Julie and Michelle, I take a deep breath before my four-mile hike down to the shuttle. I rub some arnica leaves on my tight hamstrings and descend. On the way, I practice the suspension of worry again. I try to hide my fear. But I search the horizon (and every tree) for predators. I talk to other hikers in a loud voice, as if hard of hearing. At every dense thicket, I clap my hands. At least for the first mile. Then, I forget the precautions.

Part of the serendipity of hiking in Glacier is that you discover you can't hold onto fear and wonder at the same time. So, despite pending violence, I am inevitably distracted by the overwhelming scenery, the rare beauty that sparks an immediate passion. Across McDonald Creek Valley are the snowfields of Heavens Peak and the sawtooth ridges of the Livingston Range. The raw power of the place hits me once more. It's hard to imagine this whole world of glaciers disappearing

I am in awe of both mountains and grizzlies. They stand at the apex of the alpine ecosystem, and this, for their sheer vigor and tamelessness, brings to me admiration, humility, and respect. If lions symbolize the African savannah, then grizzlies define the Rocky Mountains. Not just for their survival in pockets like

Glacier and Yellowstone, but because they once ranged farther afield, holding dominion over half the continent, like the glaciers of their mountain home. They both embody the spirit of wilderness, and in wildness, Thoreau once said, is "the preservation of the world."

A Thousand Words

The mountain landscape ahead of us does not live up to the old photograph. Usually, it's the other way around. In person, mountains and glaciers typically take the prize. Snapshots miss the mark. But this historic photograph, taken in 1936, shows a winning icescape on Kodak paper: the former Grinnell Glacier overflowing with ice. Lisa McKeon holds a copy in her left hand. In her right is an expensive Nikon digital camera. Before her is the expanse of Mt. Gould and the present-day glacier, reduced to one-third its original size. She's trying to duplicate the old photograph, if not its majesty at least its framing, in the viewfinder of her camera. So far, the landmarks do not match up. The background—Mt. Gould—looks good, but the foreground is all askew. The melt of the long Indian Summer has exposed rocks that have led Lisa astray. It's as if someone moved the ice to another neighborhood, a different backyard.

"Hey, Dan," she calls to Fagre, who is bobbing back and forth in a boulder field nearby. "A little more over here." In agreement, Fagre waves his copy of the classic photograph at her.

It is Labor Day Monday, but everyone is working. Fagre is searching in tandem with Lisa for the old vantage, the original stance, what they call the "photo point." They started the repeat photography project together, in 1997, and have replicated over sixty vintage photographs on nineteen glaciers. (Thirteen of these glaciers have shown marked recession.) For each, they have had to find the exact spot where the previous photographer stood so that the image's frame is a perfect replica. "We logged many miles searching for that spot," Lisa says. "Walking around with my eye to the viewfinder, I've nearly stepped off a few cliffs." They've also made some false starts and given up on a dead trail or two. Some photographs were unredeemable—too many features had vanished. Still, against heavy odds, they have been fortunate in matching so many contemporary photographs to originals in rugged terrain. Today, they hope for another score, that this historic image by W. C. Alden will be replicated.

Alden was a USGS scientist, much like Fagre, who tracked the waxing and waning of glaciers in Montana. He took the photograph in July 1936, before there were a lake and icebergs below the terminus. (A slight pool of meltwater, however, may have collected at the foot.) At midsummer, snow still covers some of the glacier in the photograph, but the ice extent is clear. The glacier appears robust. His photograph is a window to the southeast from the rock rubble of the north moraine to the glacier, its body wide and its frozen headwall higher than it is today. The front of the glacier appears as if it were a few steps away, like a riverfront at your feet. The most distinguishing mark of the foreground is the stratified bedrock at the left, already exposed by the retreating glacier. Yet the glacier is enormous compared to what we see now.

If a third USGS photographer stopped here, in the footsteps of Alden and Fagre, another seventy-five years from now, he or she would find a rocky landscape but no ice. The glacier would have vanished like a ghost.

"I think it's over here," Fagre calls out, in a tentative voice. The photo point is elusive. A rock out of place here. A boulder there. The eye has difficulty triangulating.

"No," Lisa shouts. "About thirty feet to your left."

"Feet?" Fagre says, converting in his mind.

He turns toward Lisa and paces off ten meters, ten strides. A scientist dreams in the metric system. "Six, seven, eight . . ."

Fagre is distracted for a moment. A big horsefly bites his leg. He hardly winces though. With his ironman hiking shorts on today, he was attacked by horseflies and mosquitoes the whole hike up. It's been hot and wet all summer and it's been a banner year for pests. But Fagre is nearly immune.

Suddenly, Lisa waves her arms at him—to stop his zigzagging.

"No," Fagre says, glancing at the print. "I'm still two meters short." But he's close to the quarry.

Lisa hands him the camera.

Fagre points the Nikon ahead as if stalking big game. He scans the archive print and then his surroundings, careful not to trip on one of the rocks scattered at his feet. He stops for a second, poised to take the picture. Then he shuffles sideways for twenty feet— that is, seven meters. A misstep. He shakes his head. Patience is required for repeat photographs, a practice more common to courtrooms than to wilderness.

And now to climate hunters. Each year Dan and Lisa compare

images for each glacier to see the changes that have transpired. Their earliest archive image is from 1887 (another shot of Grinnell Glacier), a few decades after the dawn of the Industrial Revolution. Their own replica stock library comprises the last sixteen years, bringing the archival documentation up to date. Those photographs take us back to the year of Al Gore's visit to Grinnell—1997—before the lake had significant icebergs. Serendipity made repeat photography a household phrase.

The purpose of Gore's trip was to make a speech about climate change with a receding glacier as a backdrop. Grinnell Glacier was chosen since, despite the 5.5-mile hike in, it was more accessible than most. What no one knew at the time was that Grinnell had another advantage: It was the most photographed glacier of all. In preparation for the visit, USGS staff dug up several images from a bonanza in the Park archives, ran up to the glacier to repeat them, and made copies of the pairings for the press. Reporters and cameramen from major networks, newspapers, and weeklies made the climb, listened to the speech, and broadcast or printed the paired images—the healthy glacier and its sick successor. These images were the shots seen 'round the world. More than the speech, they drew attention to the crisis of climate change globally and to the predicament of glaciers right at home.

No one was more surprised than Dan Fagre, yet he was quick to respond. Seeing the promise of repeat photographs in reaching millions of people, he launched his photography project the same year, and put Lisa in charge. Early on, they recognized that paired photographs would trump other visual records—like graphs—in convincing people and politicians that dramatic changes were taking place. For several glaciers, they took a series of annual photographs and pinned them on the bulletin board in a row. Viewers responded with an "aha moment" of recognition. Glaciers were clearly disappearing. And the evidence was obvious even to audiences that were less than sympathetic.

For skeptics of climate change, a remarkable shift in attitude took place. When Fagre added the photographs to his slide presentation at one of his lectures, the hecklers fell silent. "You can't argue with these photographs," he says. "Ice vanishing, in monumental quantities, is no hoax."

Mankind's violations against nature, against our home, become clear. The USGS photo bank provides concrete evidence of this abuse, and more proof is on its way. All twenty-five glaciers will soon be profiled. A family album of ice. To this end, Fagre and McKeon have perused over 2,500 photographs in the Park's archives searching for glaciers in their prime, at the turn of the twentieth century before temperatures really began to climb.

"I love historical photographs," Lisa says. "Each has a personal story to tell." The treasure trove in the cupboards of the National Park Service includes images from legendary Montana photographers, such as T. J. Hileman, Morton Elrod, and W. C. Alden. Hileman was called "Mountain Goat" by his friends for his habit of climbing to precarious heights with a heavy box camera strapped to his back. Elrod, who was the Park's first interpretive naturalist, recorded the ice extent of several glaciers. Sometimes he chose a boulder as a reference point, like the twenty-foot-diameter behemoth just to our left—now called Elrod's Rock—that was dropped by the glacier such as a kid would drop a ball. When he photographed the boulder in 1924, with a man casually sitting on top, it was at the edge of the ice terminus; now, the frontage is no longer visible from the same vantage point. It is eight hundred yards away. In the original image, Salamander is attached to Grinnell in the distance, even though it rides solo high above on a remote cliff today.

About his Grinnell experience and the boulder that bears his name, Elrod wrote in 1926:

> Great boulders fell from the cliffs above and gradually worked through the glacier from top to bottom, where [they] ground the mountain floor, turning like a top, polishing [the] massive base as smooth as a billiard table. A few years ago this boulder was at the edge of the ice. [Now] it is out in the rubble.

Elrod and Hileman and other early photographers were the best publicists for the new Park; their images of stunning scenery and wildlife attracted visitors from around the country. Today, however, only a few of the thousands of prints and negatives are useful to the USGS for repeating. Beauty is in the eye of the beholder. Sometimes, Lisa can't get a match because the glacier has vanished or has retreated behind a moraine. A likely deterrent is also the absence of obvious features like rocks or trees in the foreground, since they must frame the shot. And then there's the perfect match, unbelievable but right there in black and white: Boulder Glacier, 1932 and 1998. Lisa and Karen Holzer, a former colleague, set out to repeat an impressive photograph from the Depression Era, in which a string of five packhorses stand in front of a robust Boulder Glacier. A period piece. The glacier teems with ice. Lisa and Karen joked, "We'll never duplicate that." They found the photo point, and just as they were setting up the tripod, a team of mules and packers sauntered into place, in the foreground, as if on cue from Hollywood casting. Aim. Focus. Shoot. A modern duplicate was born. Only the contemporary photograph is nearly devoid of ice. The pair of images are popular on a museum tour of Lisa's repeat photography, which, along with Fagre's Web site, spreads awareness of the reality of glacier loss in Montana. Pairing historical and modern photographs has made the glacier saga accessible to the public. The beauty of a repeat photograph is its visual simplicity: It doesn't require a scientist to interpret it. Good enough for the average citizen. Good enough for judge and jury.

Dan and Lisa are now homing in on their evidence. They are trying to line up a few rocks in the foreground—by eye, not by touch—as if it were a crime scene.

"Boss, that boulder has grown," Lisa says with a smile. This is an old joke between them.

"Mountains don't move," Dan responds. He looks over his shoulder. "Step back ten meters. I think we've got it."

"Gotcha."

"Yep, this is it." Fagre lifts the camera to his right eye. "Mission accomplished."

"You know what they say?" Lisa turns to me to whisper. "Whoever doesn't learn from a photo is doomed to repeat it."

Fagre takes half a dozen shots, checking the viewfinder halfway through and at the end. He hands me the camera so I can see what he's bagged. Compared to the original, what hits me immediately is the tremendous recession of ice. The glacier has pulled back like a snail retreating into its shell. The headwall and terminus seem to be touching. In front of them lies a brand-new lake. Since distances and area are foreshortened in both images, what is most impressive is the volume of ice that is missing now. The mass is cut to a fourth.

Handing the camera back to Lisa, Fagre walks toward the lake to join the rest of the crew—Lindsey, Kevin, and Erich. Meanwhile, Lisa hangs back to tell me a story. She says she has one more repeat photograph she'd like to make at Grinnell. "I need my nine-year-old daughter, Maggie, to make the hike to do it," she says. "I have an old photo from the late 1970s of me at nine or ten in front of the glacier. I'd like to capture her in that same pose while there's still ice left to shoot."

At last estimate, Fagre gave the window for that photograph of Maggie as being the next nineteen years—through 2030. After that, Grinnell would likely be only a patch of snow and stagnant ice. There will be no photograph to take.

Six of us gather on the old luncheon rock at the edge of the lake. In four years, the exposed limestone has widened as the ice has peeled back. We would now have room for double our number, if need be. Everybody drops a backpack, each a heavy load, especially Kevin's and Erich's. They dump the two inflatable kayaks onto the slab. The Indian Summer has brought us a melted lake, ready to be navigated. Immediately, Kevin starts inflating his kayak with a hand pump. Everyone sits down, everyone except Fagre, who is perched on one foot like a crane, swatting at another horsefly or two, maybe three—it's hard to tell. Lisa nurses two blisters, one on either heel. Bravely, she forgoes the first-aid kit and straps her feet with gray duct tape. Johnson and Johnson would cringe. Lindsey looks on and claims her "feet are on fire," too, but passes on the hardware supplies. All the rituals of repair, après hike, are in force. Next: A series of sighs erupts. Relief that the march is over. In other words, the crew is ready for lunch.

Dan unwraps his usual PB&J—this time with huckleberry jam, which is in season if you can get to the berries before the grizzlies do. Lisa inspects a vegan sandwich that looks and smells like a primordial swamp. Lindsey organizes multiple bags of leftovers—she's a newlywed and buys lunch for two. Erich eats nuts, lest he put a pound on his wiry frame. Kevin and I are the only carnivores—munching on cold cuts—with apologies for our carbon footprints.

Lindsey points out the half-moon, hanging in the sky over Salamander like a kite. The sun is bright behind us so that the moon, swimming in blue, seems ephemeral and pale as if it were made of lace. Below it, the lake is dappled with sunlight, flashing like the iridescent scales of a fish. In this harsh light, sunglasses are inadequate; we all have glacier goggles on. The lake is further punctuated by two dozen or more icebergs (double the normal number); otherwise, it is free of surface ice: a window at last for the boats. Four or five of the tall bergs show rounded ridges—they're old—whereas the vast majority are new bergs with sharp edges to their exposed ice. The lofty topsides are melting from solar radiation while the underbellies—87.5 percent of the berg—are dissolving from lake waters, which are near freezing, like a well-iced martini. Besides the icebergs, several flat ice floes have been blown by the wind to the north side of the lake. Half of these floes carry rock rubble, like scree, on their flat backs. The gray rock is destined to fall eventually to the bottom of the lake.

Calving glaciers are similar to ice avalanches. The difference is that, rather than rolling down a mountainside, the ice chunk severs from the end of a glacier and tumbles into a lake (or ocean). The rate of calving seems to depend on the depth of the receiving waters. Upper Grinnell Lake is over 180 feet deep, so it can accommodate huge icebergs at a fast rate. The faster the calves drop from the mother glacier, the faster it's disintegrating. This gives Dan some pause today because a good number of the bergs are new.

We hear a loud "crack," like distant thunder, and everyone looks up and left toward the front of the glacier. Its profile is a deep blue, but not heaving at the moment. Internal grumbling. Here, the glacier is calving but it takes days for ice to break free. In the interim, it groans and creaks and roars. Calving comes with a splash, more like a tidal wave, as each of us keeps half an eye peeled on the terminus. A wave could reach us here fast.

If calving sounds like it makes up the percussion section, then "popping" is from a wind instrument. While we finish lunch, we hear a strange "pop," repeatedly, from the direction of the icebergs. This popping comes with the burst of internal air bubbles as the ice melts. It is disconcerting, like the surprise of a balloon encountering a pin. Alaskans refer to this popping as "berg seltzer" or "sizzle." The bubbly surface ice is what gives the icebergs their white color.

The conversation wanders from icebergs to cold drinks to local bars and beers. The team is divided on the best watering hole. West Glacier Bar, Packer's Roost, and Stonefly Inn all get votes. It seems Packer's Roost is the best place to see a fistfight, while Stonefly is more low key, though not a place to find a designated driver. A straw ballot draws three winners, each with two votes. But the crew is nearly unanimous with regard to Montana microbrews. At first, the nominees are varied but emphatic: Big Sky Trout Slayer, Tamarack Yard Sale Amber (which isn't that cheap), and Harvest Moon Pigs Ass Porter. The porter has backers among the men. Lindsey favors Blackfoot River IPA. I press forward with simplicity: Golden Grizzly Ale. But when Dan speaks up, the whole chorus quiets down. "If you're talking Montana," he says, "no roster is complete without Moose Drool." This brown ale probably has the best beer label in America: a bull moose exhaling his excess saliva through a gaping mouth, leaving nothing in the title to the imagination. Fagre has nominated the ringer; all hands acquiesce.

The debate over beer substitutes for a toast before launching the boats. Dan's glaciers are a dry county, of course, but talk of reverie can be second best. In any case, the Moose Drools have it.

By now, Kevin has fully inflated one of the kayaks—surprisingly ten feet long—and is working on the second. Lindsey puts on her dry suit, which is like a wet suit but watertight and more insulated. Overtop, she buckles a blue lifejacket. She hefts her red kayak, picks up the yellow-and-blue paddle, and heads for the water. The yellow GPS transponder protrudes from her backpack like a sabre. She looks like a rainbow Ninja.

Lindsey will be the lead boat, the one who paddles to the terminus and paces off the limits of the glacier with her GPS gear. Now, Kevin joins her, wearing a green lifejacket and carrying a rescue "throw" rope, should it be needed. His boat is identical to hers and also red.

Lindsey says to me, "These kayaks are built for whitewater, but nobody ever expected them to see icebergs, especially in Montana." Everyone wishes her well.

Dan pushes Lindsey off. "Be safe," he says. "Don't take any risks." She smiles at him. What could be a bigger risk? But everyone is upbeat. It's a momentous day: the first crossing of this lake in history.

Kevin follows Lindsey in the safety boat. He will not go to the wall. Instead, he will hang back and wait with the throw rope. "Better lasso me quick, if I get into trouble," Lindsey says. Their

voices become muffled as they round the first iceberg. The reflec-
tions of their boats resemble shadow puppets on the opaque ice.
The water is calm. There's hardly a wake.

The pair paddle out of sight swiftly, their eyes on the terminus.
It's a sprint to the wall, to get as close as possible and to minimize
exposure to danger. Meanwhile, on shore, Dan frets, listening to
the sounds of ice calving: *"creeek-grooor-creeek,"* like the tune of ice
trays cracking open—only on a huge scale. On the lake, Lindsey
and Kevin not only contend with the threat of calving and falling
ice but with the pernicious popping that signals melting. The
noise reminds them that melting might prompt a top-heavy iceberg to flip. The upending of a berg or floe would also bring a
small tidal wave over the boats.

"I erased my fear," Lindsey told me later. "I was nervous on
shore, then calm in the boat. I panicked after it was all over."

Navigation is tricky. They paddle into a cul-de-sac, a lead that
seizes up where two floes and a berg have bunched together. Another dead end forces them to backtrack toward us. But in short
order they reach the blue wall. Lindsey is impressed. "From the
ground," she says later, "the glacier today looks small, but when
we reached the terminus, from the kayak the wall looked immense." In actuality, it was less than twenty feet (six meters) tall.

Paddling right next to, and parallel to, the wall, Lindsey reels
off multiple GPS transmissions to the satellites. By sending pulses
a few times each second, she clocks over two thousand signals in
ten minutes' time. This will produce thousands of points to plot a
line that approximates the frontage. The life-size error will be
plus or minus 10 cm (4 inches). Thus, in mapping, a nearly exact
image of the terminus—at fractions of an inch—becomes possible. The USGS team can now track the retreat of the glacier with

greater precision. Measuring the waterfront beats surveying from atop the glacier hands down.

Lindsey and Kevin return in under forty-five minutes—without a scratch, dry as a bone. I expect them to grab a towel right away, but they surprise me and the rest of the team. On a count of three, they jump into the freezing water. What a way to celebrate. Lindsey's face turns pale blue. But it's another first: Nobody has ever swum in Upper Grinnell Lake before. The ice floes bob next to Kevin and her like corks.

Dan quits his nail-biting. "Well, now, you both are in the grizzly bear club," he says. "I guess we didn't need the boats."

We split up the team, like a posse tracking two leads. Lisa and Lindsey head north to the terminal moraine for some more forensic photography. Fagre turns south, toward the glacier, with Kevin, Erich, and me in tow. We step over the stratified bedrock of Alden's vintage photograph, now more exposed since the ice has pulled back a quarter of a mile in seventy-five years. Scattered here and there are numerous stromatoliths, the billion-year-old fossils that cling to the bedrock. We climb up an escarpment and reach a rock bench that I recall was covered by ice three summers before. I ask Fagre about it.

"The rock under our feet hasn't seen sunshine since before the Little Ice Age—since 1400 anyway," he says. Millennia to centuries to decades. This is the story of Grinnell: the contraction of time. "Just in the twenty years I've been coming up here," says Fagre, "the rock has become exposed far more than the previous hundred years."

This sort of exposure may decrease the albedo at Grinnell, the tendency of light surfaces (like snow) to reflect sunlight. Dark surfaces like rock, on the other hand, absorb solar radiation and warm up more, thereby encouraging melting of nearby ice. In a reinforcing loop, melting creates bare rock that brings on even more melting. Not only is dark rock revealed at the edges of the glacier but the gray limestone at the headwall is more exposed as well. And Salamander is framed by widening rock surfaces on all sides like an enlarging border to a portrait. Repeat photographs show the expansion over the years. Today, the heat radiates off the rock walls of the cirque like a giant kiln.

After a ten-minute hike around the lake, we reach the glacier. While Fagre and his team formerly had to climb up onto the ice, the surface is now at our kneecaps. We simply step up one flight. The ice is slim. It becomes thinner most every year. "It's as thin as a dinner plate in places," says Fagre. As we move up to the left, to the lateral margin, I see what he means: In profile, the ice resembles delicate china.

The edge of the glacier has pulled away from the rocks, forming a "moat." We jump back and forth from glacier to land to ice, over the moat, as if leaping from dock to departing ship. In this way, we zigzag up the glacier to a rock outcrop that is a reference point for Fagre. (But all the boulders once atop the glacier have slipped into a crevasse.) From this vantage, he says, "To my eye, the lateral margin has retreated some in a year, but let's see." He directs Kevin and Erich to measure the margin with their GPS gear. This information will be coupled with Lindsey's data to assess the entire glacier. Back at the lab, the lateral margin will show a modest recession, mostly from this hot summer. The major retreat is the foot withdrawing steadily from the lake.

This retraction has ranged up to 1,083 feet (from 8 to 330

meters) over ten years, resulting in the entire glacier shrinking by about 3.1 acres per year. That's a loss of 2.2 percent annually—a steady trickling away.

On a hot summer day, like today, the surface melt of snow and firn on some glaciers has been measured at 10 centimeters (4 inches). If this figure holds for Grinnell, the amount of lost water in a day could fill about eighteen Olympic-sized swimming pools. This estimate addresses only snow and firn; it doesn't count melting ice. Ice is harder to quantify, but may double the output. Ice is more dense than snow or firn, so under similar heating (say, on a warm, overcast day) it would diminish more slowly. But gray ice absorbs more solar radiation than white snow, so ice could actually melt faster under a fierce sun.

The evidence slips away underneath our feet. Spaced a few meters apart, rivulets or channels drain the glacier surface, creating a tributary network that races toward the lake and the outflow stream. Fagre points this out with a sweep of his hand. Water also percolates through the snowpack, he says, causing dimples in the surface here and there. When the meltwater descends and reaches ice, its progress halts, since ice is impermeable to air and water. Instead, it flows downhill atop the ice until it reaches a crevasse, crack, or connecting chamber. Then it descends through an intricate and poorly understood plumbing system, until it finally joins the thin layer of water underneath a glacier that helps ice flow over the bedrock. The water is a lubricant.

"All that meltwater adds up," says Fagre, pointing to another rivulet. "A huge amount of ice can disappear in 125 years or so. Just look what George Bird Grinnell claimed."

Grinnell, explorer and journalist, visited the glacier in 1887 that would bear his name. About his discovery, he said: "The glacier is a thousand feet high and two miles wide." Grinnell is a hero to Fagre but he is the first to admit his mentor might have exaggerated.

"Old George Bird was using literary license," Fagre says. "He was conveying an impression, not reporting a fact. He probably doubled the dimensions to win over his readers, I suspect. But even factoring that in, the change today is immense."

Fagre excuses the explorer and only wishes to have an audience with him. "I'd like to have a time machine so I could spend a day with that man." Of course, Fagre would have to tell Grinnell the bad news: Only around ten percent of his glacier's mass is left.

If you listen, the popping of ice echoes through these hills. George Bird Grinnell enjoyed mostly silence. Dan and I talk about the loss of glaciers across the Rockies and its gravity—the scale of the injury. "It's like the tropics losing all their trees," he says. "What would the Amazon be without its rain forest? What will the Rockies be without their ice?"

When I look above us at Mt. Gould and the Garden Wall, I can't help but think of Mt. St. Helens (formerly 9,677 feet; 2,949 meters), another white lady who has lost most of her ice. I climbed the old summit in 1974, six years before the top 1,300 feet (396 meters) of her blew into the stratosphere. I share the adventure with Dan. At the time, Mt. St. Helens was a perfect conical peak, similar to Mt. Fuji in Japan. A meditative mountain, from afar and up close.

Three of us roped up and cramponed to the summit. It was an easy ascent, zigzagging up the slope, with only one large crevasse, ten feet across and a drop of eight feet to the lower side. The two of us leapt over unroped, in the style of Butch Cassidy and the Sundance Kid. The summit snow was melting in spots due to heat from the then-dormant volcano. The odor of sulphur, like rotten eggs, was so strong we felt queasy at the top. The mountain would not be dormant for long.

On our climb, the glaciers began at the summit and ran more than 3,000 feet (914 meters) down the flanks. When we glissaded down the ice and snow, we had to break our speed with ice axes, or else spin out of control. It took five hours to ascend the mountain and one to get down.

On May 18, 1980, the volcano awakened, blowing two-thirds of a cubic mile of ash and rock into the air, with the force of twenty-four megatons. The blast packed a primal punch, carving a crater 2,084 feet (635 meters) deep and leveling four billion board feet of timber, enough to build 300,000 two-bedroom homes. The vertical plume of ash reached 80,000 feet (24,380 meters) in fifteen minutes and encircled the earth in fifteen days. The glaciers I had just climbed vaporized or were melted. The new lower summit at 8,363 feet (2,549 meters) was jagged and black. Rock, not snow and ice.

After the blowout, I viewed the amputated peak. She was cut off at the shoulders. The little lady was no longer a perfect white dome. Today, Mt. St. Helens is no longer the mountain she once was, but I think of her—she lives on in my memory of youth: a mountain of dreams. Here before me, Mt. Gould and her three glaciers look nearly abandoned. Only a vestige of ice skirts the mountain. Will Gould and these glaciers be retired to the pantheon, too—my mountains of dreams? Dan commiserates with my nostalgia. So few pockets of ice, their names nearly forgotten. For Fagre and company, it's getting lonely at the top.

Kevin and Erich are now at their high point on the glacier— just below the headwall crevasses. Without much snow cover, the fissures are obvious and present little danger. The two men move about unroped. Above them, the snow line snakes its way across the upper face of the glacier.

"It's about fifty-five percent snow and firn," Fagre says, just eyeballing it. Just below the sixty percent threshold, the ebbing snow line suggests the glacier will lose ice this year. September has been warm. The upper snow cover resembles a white lab frock, short and out of fashion, draped above a charcoal-gray skirt.

"I'm thinking about painting this glacier the way George Bird Grinnell would have seen it," Dan says, almost wistfully. "In all its splendor. Maybe when I retire I'll paint the remnant ice, too. A second portrait. But the first one is on my bucket list." We exchange glances. "Yep," he continues, "I'm looking forward to sipping a little wine and dipping a paintbrush from time to time."

When Fagre talks about painting or his retirement, it is not a good sign for the glaciers. His latest prognosis for the ice must be poor.

Fagre is near sixty now. If he keeps working slightly past the standard retirement age, he may outlast the glaciers. But not by much. Fagre seems to accept this outcome. At last, he is at peace with it.

We descend the glacier and regain the lakeshore. Fagre wants to check out the floating ice. "Several recent calving events have produced these icebergs," he says, pointing out the young bergs whose clean, honed edges glisten in the light like blades. Behind them, a section of the glacier has simply collapsed into the lake. "The two combined—calving and collapse—show a rapid break-up of the glacier," Fagre continues, moving to the far end of the lake, toward the terminal moraine. The collapse and the calving are obvious there, too. It's as though a car garage broke loose, careened down the drive, and the house roof caved in, all at once.

I assume he wouldn't want to paint that scene. It is not surprising Fagre wishes to live in the nineteenth century or even the Pleistocene. We all long for a more robust time for the glacier. After all, a clean bill of health would work both ways. If Grinnell were in good shape, we'd be pretty sure the planet would be sound, too. As go the glaciers, so goes the Earth.

When I think of the speed of ice traveling down the glacier, the scale of calving and collapse becomes even more impressive. Though never precisely measured, Grinnell's ice probably moves at one to two inches per day in summer, or a maximum (at summer rates) of thirty to sixty feet each year. Thus, it takes about fifty years or so for ice to travel the nearly quarter-mile length of the glacier. Glaciers move by gravity. At the terminus, the uphill ice pushes each potential calf toward the lake. Progress is by inches until tons of ice fall at once, and an iceberg is born.

We take for granted that glaciers march along, under their own steam, but this belief wasn't always universally held. As late as the mid-nineteenth century, geographers considered glaciers to be a mere curiosity, inert accumulations of snow. The biblical Great Flood was believed responsible for moving boulders and carving rock. Slowly, scientists accepted evidence to the contrary. First, in the 1870s, John Tyndall calculated that, if snow falling on Mont Blanc did not travel downhill, then the snowfields would become higher than the mountain's summit. Next, Tyndall and others pointed to moraines as another clue that glaciers plowed the earth. Crevasses were also offered as proof of movement, since a few men drew the parallel to molten rock, which cracked as it flowed. But the most spectacular justification began in 1820 and took forty-one years to unfold. Three alpine guides were caught in an avalanche on Mont Blanc that year and forced into a crevasse at the head of the Grand Plateau, which sits above several glaciers. The fissure filled with snow and the guides were declared lost. Decades passed. The glacier pushed downward like a winepress. On August 12, 1861, another guide discovered three human skulls at the foot of the Bossons Glacier, along with clues to their identity: two hats (one straw, one felt), a pigeon's wing (one of the men had planned to release a pigeon or two at the summit), and a boiled leg of mutton. The skulls and accoutrements had traveled 9,250 feet (2,819 meters) down the mountain, or approximately 226 feet (69 meters) per year. Glaciers did, in fact, move downhill.

As ice flows and melts, history is revealed.

Today, skeptics doubt that glaciers retreat *uphill*, or at least doubt why they do so. This time they can't blame it on the Great Flood, but many—in dismissing glacier recession—claim that the planet is actually cooling. Thousands of thermometers around the globe must be faulty. That's one explanation. Yet it's simple grade-school science that heat melts ice. Greater warmth or less snowfall are responsible for ebbing glaciers—exactly the predictions of climate change.

Lisa and Lindsey wave frantically at us from halfway up the terminal moraine. They have a hundred feet left to go. The hill is beveled: broad, perhaps 250 feet (76.2 meters) wide at the base, and a slim 3 feet (0.9 meters) at the top. The narrow ridgeline runs for half a mile. Dan and I are walking northwest of the lake, past Elrod's Rock when we spot them. Dwarfed by the huge embankment, the pair appears to resemble children. The moraine is daunting, with treacherous footing. Lisa yells, warning Lindsey of rockfall stirred by her boots. They are climbing to get a bird's-eye view of the glacier to put a new technology into play: gigapixel photography. It's the hot new thing. I saw this imagery on Lisa's computer screen back at the lab and was mesmerized.

Gigapixel photography is a new edition of high-resolution imagery, boasting more than a thousand times the information captured in a regular digital camera. Scientific uses range from archeology to space exploration. The technology was originally created for the Mars Rover. The leading model, produced by GigaPan Systems, comprises a robotic camera mount that controls the panning and shutter release during a panoramic shoot. A microprocessor calculates the number and size of the exposures to assure enough overlap of images to create a complete and seamless wide view. Back at the lab, as many as eight hundred photographs are stitched together with post-production software into a single, merged image. All this is cutting-edge photography, but even more innovative is the viewing software. The Web audience can zoom in on any part of the panorama and bring the most remote objects into focus. A mountain goat or an iceberg can be pulled into close view. For glacier hunters, this is the jackpot: Forensic documentation is possible in enhanced, almost unlimited, detail.

By the time Dan and I hit the base of the moraine, Lisa and Lindsey are on top, nearly two hundred feet above. While they set up the tripod and program the microprocessor, we trudge up the slope. The ascent is a challenge, since the rubble gives way under foot, similar to the poor purchase one experiences when climbing a sand dune. It's difficult to get any traction, and most footholds are false promises that go hurtling past and below.

We attain the crest. There is hardly room for four. Lisa stands next to the camera and replaces the batteries—eight double-As. Lindsey stoops below, clutching the tripod, bracing for any gusts. I relieve her so she can duck out of the wind. She is thankful, still half-frozen from her swim in the lake. While I hold the tripod steady, Lisa sets the levels to keep the camera plumb.

"We're good," she announces. "We have 345 shots to make—it'll take an hour." (Stitching the photos together will take another five hours on the computer back at the lab.)

The camera begins its arc, a counterclockwise path from behind us, to Salamander, the perfect amphibian-looking glacier at our right, then across the lake and up to the main glacier at our left. The camera's "click" every half second or so is surprisingly comforting, a familiar sound at a remote outpost in the eye of a brewing storm. Cold and shivering now, we don wool hats and gloves, and huddle around the tripod like a human teepee, a windbreak for the lucky few.

Dan talks about the 160-year-old moraine with intimacy, since it is his own backyard turf. "It took centuries to plow the earth into this mound," he says. "One day around 1850, the glacier arrived here and dropped its last load. Then it recoiled. Humans have been looking over their shoulders ever since." It would take hundreds, he says, perhaps thousands of dump trucks to return this rubble to the basin uphill. There the glacier no longer plows a moraine; it just drops the occasional boulder, as if from exhaustion, as it retreats. Eventually, only a pocket of ice will sit at the headwall. The trail to the glacier's source, its final resting place, will be littered with a string of rocks, like a series of periods, or ellipses, at the end of an incomplete sentence, not coming to the point.

The idea of a glacier, he continues, is that it is an integral piece of the ecological puzzle, a reservoir, an oasis to wildlife, watersheds, and humanity. But Grinnell's legacy is no longer to be ice, not frozen storage of water for the benefit of all. Rather, its legacy is to be meltwater in a lake, soon to be a "tarn," or glacial pond. The glacier will then retreat out of touch from the pool, which will be fed only by snowpack and rain.

"Are we talking one decade?" I ask. "Two decades? Fifty years?"

"Let's talk about it on the way down," Fagre says, gathering his thoughts or perhaps delaying his confession.

Lisa and Lindsey divide the equipment and place it in their packs. Lisa says, "We'll repeat even this GigaPan in a few years to see what time shows." Each of us heads down a slightly different route, to protect the others from rockfall. I have the benefit of a game trail to guide me off the moraine. I'm parallel to Fagre when the headwall comes back into view. It is impressive: all that ice. He stops in his tracks, and I come alongside.

I half expect him to speak of painting again, but his talk of brushstrokes has grown silent. He's thinking even farther into the future now, about something that's in everyone's thoughts. It is ironic in a place of rock and ice so steeped in history, in the geological past, that we have tomorrow etched so sharply on our minds.

"Grinnell is probably deteriorating faster than all six other glaciers that we study," Fagre says, finally. "That's the cost of calving into the lake." We are on the leeward side of the moraine, so we are free of the wind and the cold. Fagre is just getting warmed up.

"The old computer model," he continues, "the one that predicted the Blackfoot/Jackson Glacier Complex would disappear by 2030, also estimated it would whittle down to 0.94 square miles (2.44 square kilometers) in 2010. But the glacier actually came within striking distance of that in 1998." That's nearly twelve years early. This suggests glaciers were being reduced at least a decade earlier than predicted. It looks like the big melt is on.

I mention that Myrna Hall, the co-producer of the Blackfoot Model, recently compared temperature projections for 1990–2007 and discovered actual temperature increase to be twice the model-predicted rate.

"That's right," Fagre says, kicking at a rock. "Without a meaningful reversal in the upward temperature trend, functioning glaciers will disappear, perhaps as early as 2020. Not 2030: a decade earlier than what we thought." He pauses a minute, then adds, "Heavy winters could push that timeline back a few years, but reckoning day will likely come sooner rather than later."

Conservative, as always, in his forecast, Fagre's words jolt me. At worst, the glaciers have nine years left. I'm stunned. It's like hearing a good friend has been given a death sentence.

We continue our walk—really, a stumble—down the moraine. I kick a couple of rocks loose, and they roll toward Kevin and Erich below. I yell, "Rock!" but my voice is muffled by the lingering wind. Like good mountaineers, the young men see the rocks coming. Other warning signs—the sight of smoke-like puffs of dust—beat my holler to the bottom of the hill.

I drop my head—my eyes more critical, my feet more careful—to continue the descent. I make a cautious step or two. But Fagre suddenly cuffs me. "It doesn't look now that Blackfoot will be the last glacier in the Park," he confides. "The field is jockeying for position. Harrison is the biggest now; Blackfoot is only second. And there's more to it than acreage. Southern exposure would put any glacier on a faster track.

"If I had to bet, I'd choose Weasel Collar as the last glacier to survive. It's isolated, partially secluded from the sun, and generous of depth. The cirque is a deep pocket, creating its own icebox. So it's self-preserving. It shows the lowest percentage drop in area of all the glaciers for the past forty years. A cool little glacier. It just could be the one."

Weasel Collar: It sounds like a Montana microbrew. But the glacier is real enough. It is nestled on the northeast side of Mt. Carter and registered at 136 acres (0.55 square kilometers) in 2005, tenth in a field of twenty-five. It will have to come from behind to win the race: Weasel Collar by a nose.

Whichever one is the last to go, it won't die in obscurity. I imagine that the last viable shelf of ice will become a household name. Harrison, Grinnell, Sperry, Weasel Collar, or another ice field will be tagged as the last glacier of the national park that bears its collective name. Many people won't notice its death throes until it's gone. But, after its demise, everyone will wonder where the glacier went. And why.

Right now, Fagre seems disappointed by his own news—the likelihood of a foreshortened timeline. Still, he is resigned to the ultimate consequences of a park without glaciers. He has placed his optimism elsewhere: the chance of saving glaciers outside the continental United States. "The lesson from Montana," he says, "is that only swift action can save the world's ice."

The men gather at the base of the moraine and wait for Lisa and Lindsey. We hover at the junction between the spur to the glacier observation point and the long leg down to the parking lot. I pull out the trail map, just to get oriented. A young woman—about twenty-five and striking—kicks up the trail from the cars, leaving her boyfriend in the dust. She asks me directions, and I point toward the lake spur, but she seems unsatisfied. She asks for a peek at my map.

I nod. "Grinnell Glacier appears larger on the map than in reality," I say, passing it to her unfolded.

"How far?"

"Half a mile maybe."

"Good. I came here to see as many glaciers as we could, before they disappear."

Her companion now wanders into view and, with just a wink, grabs the map and plops down on a nearby boulder, a rock dropped by the glacier perhaps when the Civil War was underway. I tell them Grinnell carried the boulder for half a mile on her back.

"All this history blows your mind," the young woman says. "And it's all melting. Just like that. I guess we're lucky to see Grinnell at all. I'm sad for the glacier but I'm ecstatic to have made it this far—all the way from Michigan."

The angels are not always on the side of the glaciers, however. I ask Fagre about the climate deniers he encounters on his lecture turf, the exact opposite in philosophy from the Michigan couple. The climate skeptics show up at his talks occasionally, with derision, just like creationists heckling a school science book committee. While some are stopped dead in their tracks by his repeat slide show, a few others are persuaded by his oral arguments. He doesn't make the link to whether humans are causing climate change; rather, he sticks to the impacts of warming he's been witnessing on the ground in Montana. He simply demonstrates that glaciers are waning. Facts and figures. No conjecturing. The presentation is bulletproof.

A frequent refrain from the remaining skeptics is why worry

about a little warming when it's soon to be cooling. They claim another ice age is right around the corner.

"They forget the difference in time scale between an ice age and abrupt climate change," says Fagre, setting his backpack on an aspen stump. He bats at a fly but misses. "The massive changes we're seeing around the planet are happening over just a few decades. Ice ages happen on the scale of millennia. The next glacial period may be two thousand to five thousand years away or more. But the arctic summer sea ice may be gone in ten years. Which do you think we should be concerned about? One is a crisis today. The other is so far in the future you might as well be asking what we should do when the sun quits burning."

Fagre swats a particularly large horsefly on his leg, which falls like a dead soldier. Then, he adjusts his shoulder straps and re-hoists his generous pack onto his back. It's not quite full; his two mules today—Kevin and Erich—are carrying the deflated kayaks. They strap down the boats with parachute cord and lift the packs onto their shoulders. We are set to go.

But Fagre has more to say. "The problem is not just the skeptics or the deniers; there are millions who recognize global warming but are reticent to stop it. The climate issue is just too vast, too large for some to grasp. It's intimidating. It's a challenge to get your head around it. But more than that, really, it's basic human nature to only respond to a crisis. If the Columbia River or the Colorado gradually loses its water because of lowered snowpack, few people complain. But if, suddenly, Phoenix has to be evacuated because there's no air-conditioning, electricity, or water at all, then people will take climate change seriously. They'll jump up and demand action. It's like the old parable of the frogs and the pot of water. Drop a live frog into boiling water and he hops out quickly. But if you put some frogs into cool water and slowly bring it to a boil, they all cook—they die."

The five of us exchange glances. The young hiker smiles and joins her companion on the boulder, where they inspect the map more closely. Meanwhile, Lisa and Lindsey descend the moraine by rushing down another goat trail to catch up with us at the junction. By the time they arrive, Dan and I are in the middle of another discussion about the peculiarities of human nature.

"It takes a generation or two to accept a radical idea," he says, scanning his crew. "Look at the history of science: Copernicus, Darwin, Einstein—they've all had their naysayers. Next, people didn't believe in continental drift; they thought the Earth was solid, the surface immovable. Now we have climate change. The valid theories have proven true over time."

"But why a generation or two?" I ask of the crowd. "Do we have that kind of time?"

"Children and grandchildren will correct our mistakes," someone says.

"They may blame us for what we've done," says another.

Lindsey pipes up: "That's our best bet—the next generation. Young kids hear about global warming, about the loss of glaciers, in school these days. We never did. They'll care more than we do, when it's their turn."

"What's happening is very scary for kids," Lisa says. "And the worst is yet to come. It's sadder for what they'll witness than for me. My oldest daughter—Maggie, just nine—told me she couldn't sleep the other night because she was worried about things changing so much."

Dan is deferential in his response. "I hear you. But when your girls get to the age of mine—in their twenties—it won't be as hard for them to get their heads around climate change as it is for us today—young and old. That's why it takes a generation. My daughters view climate change as part of the common dialogue. They grew up with it. Young adults now accept it as real."

"Some do," adds Lindsey. "But for others, the impacts are too remote. The loss of glaciers is maybe more tangible to us, living and working here in the Park, yet even we can't totally grasp the glaciers being gone. I can't. It will change everything. Alpine wilderness won't ever be the same."

"Just explaining that loss is overwhelming for kids," Lisa says. "But, despite their fears, my girls will come of age somehow. Then they'll have the chance to adapt to all the changes. Their success or not will brand this century. That's the future right there."

The six of us shift the pack weight on our shoulders and hips, getting ready for the long descent. Dan stoops to spray some bug repellent on his bare thighs. A reluctant concession to modernity. I slap my shirt pocket and jeans, an old habit, locating my compass and good-luck stones. What's missing, of course, is my trail map. The young lady wandered off with it.

"I've lost my map," I say to Lindsey, the closest crewmember nearby.

"That's all right," she says. "I know the way down: You just follow the meltwater—rivulet to creek to river—on its way to the burgeoning sea." She smiles at her own grandiloquence and heads down the trail, first among equals as always.

Yes, I guess I know the way down, too. I can name the landmarks. From "rivulet" to "moraine" to "tarn," I can now speak "glacier" fluently. Mastering this tongue is bittersweet, however. I realize the alpine dialect, full of cirques, horns, arêtes, and bergshrunds, will soon be a forgotten language. In Montana, at least.

And the ripples of reckless extinction will fan out, each a small wave, claiming mountain range after mountain range, until the vernacular is gone. The names of the great glaciers will be lost to our grandchildren.

Our descendents won't look upward in awe at the ice, or drink of its cool waters. They won't catch the native fish or easily watch a wolverine forage at the foot of an avalanche. Likely, they won't see the firn line on a glacier, or speak of it. It will just be too warm.

Lisa, Dan, and I take up the tail end of our single-file descent. In that order. I like to be last, so I can occasionally stop and take in the grandeur. Lisa calls out to me over her shoulder: "Around this bend is the last chance to see Grinnell. Will you be coming back, Chris?"

"Just a minute," I say. "Let me take a look around the corner."

"Outliving these glaciers is a little like outliving our children," Dan says, stepping aside to let me by. "You pray it never happens."

"The time to save Grinnell was thirty years ago," says Lisa, backing away on the trail.

"Unfortunately, we didn't take enough precautions this time around," says Dan. "That's the tragedy: There's no saving Montana's glaciers. But there's hope elsewhere—the Alps, maybe, and the Himalayas—if we learn from our mistakes."

We round the bend. I drop my pack and pull out my camera. Then I turn around. Above me, the triptych stretches wide: Gem, Grinnell, Salamander. All glistening white and gray like diamonds in a platinum setting. A photograph can't capture this beauty: Only by hiking here, five miles up the trail, by drinking glacier water and eating huckleberries, by watching a grizzly, can you appreciate its essence. I put my camera away.

Hoisting my backpack again, I get one final view of the glaciers. They remind me for a moment of a cloudscape, where animal figures seem to materialize out of the mist. To me, Salamander chasing after Gem makes them look like eponymous characters in a fable. Grinnell looks like a thin child reading in bed. A frail child, I think.

I ease forward to answer Lisa's question. "Yeah, I'll return in a few years," I call out, hoping the patient may revive.

"A few years may be too late," Lisa says.

"Yeah," says Fagre. "Unless this hill freezes over." He poises on one toe and kicks a pebble down the trail.

YEAR FIVE: EPILOGUE

The Widening Gyre

It's summer again, and a scorcher. Wherever you live in the country, the heat is turned up. Montana is no different, as I learn today—a sweltering July afternoon. After five years tracking the glaciers and their surroundings, I've come full circle: I find myself at the Park entrance again, in greater heat, asking many questions of why and when. Among them: Why have we turned our backs on the glaciers and when will we embrace a solution?

I stand at the edge of Lake McDonald, that nine-mile-long trough cut by the now-extinct McDonald Glacier. The ice field was once a mile thick, here at its trunk. A warming Pleistocene climate, a natural cycle, spelled doom for this behemoth; now modern warming sets the countdown for Sperry and twenty-four others in the Crown. Today, the offender is humankind. Scientists largely agree on this. Their verdict is on my mind.

Who gave us permission to seize control of the climate? Like an invading army divvying up the spoils, we just took it. The atmosphere is just the latest to fall under this offensive. We have usurped land and water resources as well. The dilemma for all of us now is a boomerang effect: There are consequences to our seizure, to our tampering with the climate. Our profligacy is coming back to haunt us.

Civilization, from the farm to the hydroelectric dam to the city, has become dependent on the global layout of ice. As that ice dissipates, we will see the rising sea water and drying watersheds. We are not prepared for that sort of invasion of the elements. Or casualties. People will suffer, especially in the mountains and deserts and along the coasts. Much of our mountain water storage will vanish. In the Indian subcontinent, for example, farmers and livestock will wander or die. Over fifteen million climate refugees were counted worldwide in 2010. Ten times that number are projected by some forecasters for 2050.

The modern warming also brings speed of destruction— thousands of years are compressed to twenty or so in Montana. To one hundred or so elsewhere. For the first time in perhaps seven thousand years, Montana will be nearly barren—naked of any significant ice.

I've returned to Glacier National Park for a last look at the glaciers of Montana. They've had a bad year: modest snow and a steady summer melt. A net loss of ice is likely. Nationwide, the year 2012 was the warmest year on record. Fagre's team will be surveying the glaciers once more to tighten the forecast for the ice. I think of that cool ice water peeling off the toe of each glacier, trickling into streams, and joining creeks and rivers in its flight to the sea. The wastage runs all summer long. This big thaw at Glacier is only the beginning. The pulse is matched and magnified around the world. Think of the glaciers of an equivalent thirty thousand Glacier National Parks: the meltwater from Greenland, the Arctic, Antarctica, and mountain ranges on five continents, all causing the oceans to rise. The seas already nip at our heels. The marshlands around New Orleans, for one, are eroding. Entire coastlines are now in harm's way. Like the days of the first explorers, the map of the world will change.

Standing on the lakeshore, my mind goes back to another hot summer day, a sizzler for the record books: the North Atlantic, 1983. My partner, Donna, and I were sailing on the *Eagle*, a 105-foot yacht, between Nova Scotia and Newfoundland. On a whim, we headed southwest from St. Pierre Island to Sable Island, a fifty-mile-long half-moon of sand, situated 190 miles (305 kilometers) southeast of Halifax. During the last ice age, when oceans were lower, the sand spit was likely a barrier island or a point of land. Or perhaps a huge terminal moraine from a massive Canadian glacier. Today, the offshore island is only manned by a weather-station crew but has had its fair share of rogues and pirates in the past. It is surrounded by treacherous surf and by hundreds of shipwrecks. No doubt bullion lies below. Divers have never excavated the ships because they are patrolled by great whites and other sharks that feast on the island's seal population (which in turn forages on cod). If this obstacle wasn't enough to visiting Sable, the Canadian government has placed the island off-limits; special permission and a permit are required to land there—by helicopter or boat. Preservation at its purest. We had an edge, however; one of our shipmates was author Bruce Armstrong, who had just written a history of Sable; he secured the necessary permission over the radio. We anchored off the north side of the island, the first private vessel to do so in twenty years.

We planed over the surf in our Zodiac dinghies and hit the beach like marines landing at daybreak. Despite the sea breeze, the temperature was over 80 degrees F (26.7 degrees C). I had only recently heard of the theory of global warming from a professor at the University of Maryland. The heat in the North Atlantic seemed like an aberration.

Scattered on the beach were sunning harbor and gray seals and their pups, which were unfazed by our presence. They had not yet learned to fear man. Beyond the rookery, the sand spit stretched unbroken. It was littered with whale skeletons and seal bones. No collectors were allowed to remove the bleached treasure. Few shells, though. They must have been broken up in the surf, which roared and reared and broke just offshore. We explored inland—the island was about a mile wide—where we could hear both surfs, north and south, crashing on opposite shores. In the middle, sand dunes supported pastures of sea oats and heath, the only nesting site of the Ipswich sparrow in the world. And standing in the field of oats and marram grass were the biggest surprise: horses, wild horses, everywhere.

Legend has it that a Spanish galleon shipwrecked offshore one day, and before the ship was ripped up in the surf, some of its equine cargo escaped, swimming to shore. More likely, settlers or pirates brought horses to the island to graze. In any case, the wild horses—bays, chestnuts, and palominos—have inhabited Sable Island for nearly 250 years. They have a refuge, free from predators. But the population still has its limiters. In winter, the herd loses a fair percentage, especially foals, to storms. However, they've had a respite of late. The winter temperatures are more modest and the snows less frequent. The population surges and is near its carrying capacity. Something far more lethal threatens the horses now.

The seas are rising. Each year the high tides cut father up the beach. The waves carve up the wrack line. The sand returns to the sea. Sable Island is particularly vulnerable to this intrusion. It comprises a series of sand dunes, constantly shifting, sculpted by wind and wave. The island at its best is ephemeral, blowing east for a decade then west, re-creating itself from the shallows, its sickle-shaped countenance reappearing on old footings like the phases of the moon. It barely holds on to its perch in the North Atlantic, as if one stiff breeze could blow it all to sea.

A 3 foot (1 meter) sea level rise would effectively do the same. Submerge the island. That's what is expected by the twenty-second century—a hundred years from now—when the grandchildren of my grandchildren come of age.

Around the globe, low-lying islands like Sable are endangered by sea level rise. Up to 23 inches (59 centimeters) of rise in ocean elevation, on average, is expected by the end of this century. More in the decades beyond. Sable will lose its wild horses; Tuvalu, in the South Pacific, will lose a nation. The same story is repeated on island-states around the world, from the Maldives in the Indian Ocean to American Samoa in the Pacific. Plans for evacuation from these islands are under way. Climate refugees will have to search for new homes.

What does this have to do with vanishing glaciers in Montana?

A great deal, I think, as I drive around the lake toward the Trail of the Cedars, a meditative path at the base of Avalanche Creek. I picture lost ice, lost real estate, lost diversity. Climate change touches every corner of the globe and every thread in nature's quilt. First, seas rise; later, rivers run dry. Impacts penetrate deeply, like ink spilled on a bolt of cloth. Snowmelt in Glacier affects flood levels in the Cut Bank River, then downstream in the Missouri, and finally in the Mississippi. The less snow and ice, the lower the flood. The timing of flooding is also important: Earlier snowmelt has tampered with the traditional surge. Spring floods are vital to agriculture around the world.

The accelerating loss of ice in Montana also informs us of what's to come in the Alps, the Andes, and the Himalayas. Glacier, like Sable, is an early warning system for a planet that is heating up. Sum up all the coming ice melt, especially that from the poles, and we have submerged much more than Sable, Tuvalu, the Maldives, and Samoa. We will have inundated half of Florida, the Netherlands, and Bangladesh. Expect mass exodus—like that during our nation's Dust Bowl years—from those countries as well.

Even if Montana makes only a small contribution to sea level rise, the extinction of her glaciers (and lessening of snowpack) will nonetheless have an immediate impact on many other fronts. The famed water towers of the Crown will run dry for much of the year. No dependable summer reservoir of water. The Missouri River, Columbia River, and Saskatchewan River watersheds will bring less drinking and irrigation water to people downstream. Fish and wildlife will likely be displaced. Tree lines will climb, and avalanches may partially close up shop. Habitats will shift and constrict. The legendary biodiversity of the mountains might wane.

All this is hard to imagine.

We think of Glacier as a bastion of snow- and ice-clad peaks, carpeted below by meadows and forests with prolific herds and predators, with trout leaping in the streams. A place of tranquility and wildness. A source of our waters.

Yet this picture, the full spectrum of its wealth, could vanish in a breath.

Yes, warmwater rainbows and lake trout could survive but perhaps not the native cutthroat that is tied to local cold water. Yes, marmots might still sing on the talus slopes but not the pika.

Lynx, wolverines, and snowshoe hares may simply head north where snows still blow in Canada.

In Montana, forest fires will probably accelerate. One more record-breaking, hot, dry summer and a fair chunk of the Park may simply burn down. With it will go the dens and burrows and nests of countless creatures. The species profile of the mountain forest will likely shift. A big burn will be slow to recover, if at all.

Habitat loss is the leading threat to biodiversity, to endemic and endangered species, worldwide. To the list of culprits—from clear-cutting forests and drilling prairies to draining wetlands and poisoning waters—we can now add the thermometer. The ability to raise the temperature of our neighborhood, as well as the planet, is in our hands. We are now responsible for the changing climate. Do we want the responsibility? Not all the potential cascading impacts are known.

"There will be winners and losers," Dan Fagre tells me on my cell phone. He is on his way to a GLORIA conference in Scotland, where he and other climate change experts will discuss the future. I call him monthly, just to check in. "Some species will be able to adapt," he says. "Others won't. Young subalpine fir will expand their distribution while alpine meadows will shrink. If that happens, mountain goats could lose out, too. So will snow-dependent carnivores like the wolverine, which has a narrow niche. But highly adaptable animals like coyotes and foxes, which thrive in disturbed habitats, could end up as champs."

Victors and victims will dot the hillside. Many stories will be unexpected. Farther down the mountain, red cedars may benefit from the increased rain.

At Lake McDonald Lodge, I drive beyond cell-phone range and lose reception with Fagre. In another fifteen minutes, I park at

Avalanche Creek and walk to the nearby cedar grove—nearly a pure stand, with only a few scattered hemlocks. The ground and foliage are wet, like a rain forest. The enormous trees, with seven-foot-diameter trunks, reach two hundred feet upward like buttresses on a cathedral, shutting out the sky. A maritime species, these western red cedars are the easternmost stand on the continent. Inexplicably, they found a cool, wet pocket, a shallow depression with ash-laden soil, to inhabit in what would become Glacier National Park.

They are hundreds of years old, not thousands like the glaciers, but ancient nonetheless.

The Trail of the Cedars, an elevated boardwalk, takes me into the heart of the old-growth forest. This is where I will come when the glaciers are gone. The trees close in, until all I see is their shaggy reddish bark and flat, fanlike sprays of yellowish green foliage. Glancing upward, I see that the trunks rise like pillars, more or less unblemished until the top branches crown the trees. I am reminded of redwoods, but these cedars are closer packed and subdued in color, as if washed with gray ink.

The grove is deep and quiet. Closing my eyes, the only sensation is the smell of the trees' resin, the odor of cedar chests and closets. Maybe the moisture and decay of the forest floor reaches my nose, too. Few other plants grow here—the sunlight does not penetrate enough—except for young saplings of red cedar and hemlock that got their start from the tiny cones that litter the ground.

The big trees got their seed from the cones that survived the fire of 1517, which tore through this basin nearly five hundred years ago. Tree rings of a few felled trees tell this epic story: A cataclysmic fire raced through the hollow just before a major climate shift. The blaze cleared the land, making way for regeneration. At the doorstep of the Little Ice Age, the cool and wet years that followed nurtured the cedars with their optimal growing conditions, their best climate envelope. (They probably could not get started again in today's warming environment.) The oldest and largest cedars today found their roots back then, grew, and survived. No other fire has raked this forest in the five centuries since they were seedlings. (Adjacent forests have been burned in 1861 and 1913.) If fire stays away, the king cedars could live another 500 years, topping a cool 1,000.

Yet those years will not be cool. Not anymore. The Park's three monarchs—the glaciers, the grizzlies, and the cedar groves—those that typically live longest, are all in danger. The shifting climate will likely affect each in singular ways: glaciers by melting, grizzlies by shorter winter denning (and thus greater exposure to people), and cedars by the increased threat of fire. At first, these indictments appear unconnected, but climate change scribes its signature to each.

Warm and dry is the new imprint. A light winter, a brief winter, makes less snow for bear denning (and less snowpack for river flow). An arid summer furnishes dry tinder for forest fires. Both—a tepid winter and hot summer—create a negative mass balance for glaciers and a steady retreat of ice.

Cool and wet—the legendary, sustained subzero frosts and flurries of Montana—are largely a script of the past. Settlers and Native Americans knew them in the 1850s, in the last days of the Little Ice Age. Today, any big snows are often followed by warming rain. The glaciers run slick with fresh water.

I have always been at peace in the silent embrace of a glacier cirque. My mountaineering adventures have often brought moments of such tranquility. The half circle of ice and rock tends to muffle all sound. The fog over the frozen surface accentuates the hush. Today, the deep quiet of the cedar grove reminds me of that silence. The grove holds a mist, too, as if the trees were breathing. In a way, they are: gulping carbon dioxide and expelling oxygen and water vapor. The mist on glacier or grove looks like a breath— thin over here, thick over there—as if hesitating on exhale.

The greatest common factor—and by that I mean the greatest affinity they share—is their embodiment of time. Time is palpable in the grove, as it is on a glacier. The progression of years is visible at both. Downed logs show off their annual tree rings. Crevasses sport similar layers in the ice. Cedars expand their girth two to three inches or so a decade. Glaciers grow or recede at thirty or so inches per year. You can measure each. The growth is slow, so their old age is apparent. Their seniority is on display.

I feel a reverence for venerable age. The elders of a tribe exude an air of wisdom, regardless of the species. Standing here among the sentinels, I wonder how anyone could shoot a grizzly or cut down a redwood. How could anyone burn a cedar or melt a glacier?

Quite easily, it would seem.

My reverie is broken by the trill of a red squirrel, then another, chattering, which cuts the silence back and forth like a pair of banjoes dueling. Just as quickly they go mute. I scan the forest for more wildlife but only come upon a varied thrush sitting on a cedar sapling. The female looks virtually like a robin, but with an orange eye stripe, creating a charcoal cheek patch that suggests the rakish appearance of a bandit. She flitters off without a word.

The community of the grove has coexisted for centuries and, before the 1517 fire, for perhaps millennia. The trees have witnessed generations of thrushes, squirrels, bears, and people. Native

Americans used to harvest stringy bark for clothing, cradles, blankets, and rope; the entire plant had medicinal uses. This was back when the climate was in control. In those days, the trees witnessed the advance of 150 glaciers. Now the clock is not only running down; it is accelerating in its decline. The cedars may see the year 2100; the glaciers definitely will not.

Walking now deeper into the forest—I am stalking the thrush from tree to tree—I think about my mere fifty-five years, one-tenth of the age of these giants. What has my own generation witnessed? Fifty glaciers cut in half. Twenty-five now teetering. The cedars witnessed this, too. The difference is that they were witnesses only; they did no harm.

The thrush breaks the silence. She whistles a single high note for a stretch, then pauses and resumes with a similar tone—a watery *chup*—at a lower pitch. In the mist the song is reminiscent of a boat's whistle and I am enthralled. The varied thrush is a keystone species for the cedar grove and must have been here ever since the fire—her ancestors and their descendants.

All the mountain environments—forests to meadows to glaciers—share this history: the ebb and flow of the ice ages and the birth of modern climate change. That's the far and recent past. So how do the monarchs of the mountains inform us of the future?

The glaciers, for one, are speaking.

Withering ice hints at an ominous specter on the horizon: climate checkmate. At some point in the melting, climate impacts win; the damage is not reversible. Environmental checkmate may be sudden and the collateral damage swift. In the wake of a glacier's demise, the downstream players fall like pawns. The traditional aquatic life—in Glacier, stoneflies and native fish—vanishes. Such a moment of no return is sometimes called a "tipping point," but scientists commonly refer to it as an "ecological threshold." Brought on by an external force like climate, the abrupt shift in structure, diversity, and functioning of an ecosystem is often permanent. Once a threshold is crossed, the ecosystem is not likely to return to its old character. Another example of threshold behavior is the advance of drought-tolerant, woody plants into grasslands throughout arid regions of the world. This climate-induced shift has transformed perennial grassland into desert. At some point in the drought, desertification was a fait accompli—checkmate.

Climate change is now pushing more and more ecosystems toward thresholds. These tipping points are not well understood but are extremely important in nature and for human society in anticipating a changing world. Among scientists, there is increasing recognition that minor changes in climate can trigger precipitous reactions in the environment, whenever a threshold is crossed. Threshold crossings are prevalent in Alaska because, like Montana, temperatures have climbed above the global mean. Scientists are hard-pressed to explain why the heat has escalated near the poles, but feedback loops are certainly in play. Warming of Alaska has prompted various sudden effects, ranging from the reduction of sea-ice coverage to the melting of permafrost to the earlier arrival of snowmelt in the spring. Impacts have trickled down to coastal fisheries, wetlands, and forest communities. A vivid example of such a threshold crossing can be found in the Alaskan tundra.

Above the Arctic Circle, north of tree line, warmer temperatures have reduced the duration of snow cover. A more modest winter and the lack of white snow in early fall and late spring has lessened reflectivity at the tundra surface. Diminished reflectivity prompts greater absorption of sunlight, bringing on more warming, which, in turn, accelerates the loss of snowpack. The feedback effect amplifies at each turn, quickly leading to warmer conditions that nurture the invasion of shrubs into the tundra. It does not stop there. The new shrubs are dark, absorb sunlight, and further foster the warming. The net result is a rapid chain of events converting arctic tundra to shrubland. It all began with a relatively small elevation in temperature, but is ending with the propagation of a new ecosystem. In northern Alaska, shrub coverage has climbed sixteen percent since 1950. The overall Alaskan tundra story suggests a paradigm shift. And the chess pieces keep falling. Experiments have shown that when warming stimulates the growth of shrub cover, it also reduces lichen cover, an essential food of caribou. Most arctic caribou herds are dwindling, though the reasons are not fully clear.

New tipping points are emerging around the globe. In my boyhood home along the shores of the Chesapeake Bay, salt marsh islands are quickly disappearing because of sea level rise and subsidence. Several large islands, like Hollands and Sharps, that once harbored communities of watermen, have vanished under the waves. More than a dozen island ecosystems in all are not coming back, their saltmeadow hay replaced by underwater eelgrass. Consequently, marine charts of the Chesapeake are now and forever out of date. Not only islands are vulnerable. Erosion also bites at the mainland.

Threshold responses are becoming so frequent that there is no need to look beyond one's own backyard. Right at my current residence in the mountain forests of the Southwest, warmer and drier conditions are placing piñon and juniper trees under greater water stress. Threshold crossings between vegetation types are imminent here. A sustained drought—and the collateral damage anticipated from likely fire and insect outbreaks—could bring about a dramatic forest dieback. A hotter and drier Southwest— and a browner one—may lie ahead. A different ecosystem may inherit some of the land.

Change is not the problem; ecosystems are not static. Succession, for example, is always under way in fire zones and avalanche tracks. The concern among scientists is over precipitous change, as may occur in the Southwest, Far North, and some places in between. Species cannot readily adapt to a disrupted environment.

In the middle of the continent, the mountain bark beetle is decimating timber from Colorado to Alberta, the main spine of the Rockies. The crisis is most severe in the Canadian provinces: British Columbia has lost thirty-three million acres of lodgepole pine forest. Lodgepole pines are native to higher altitudes, but the beetles have now invaded ponderosa pines in Colorado's Front Range, which suggests forests may be killed around densely situated homes. Scientists worry the beetles may migrate all the way to the Great Lakes.

The infestation is tied to a threshold response of the beetles to warmer temperatures, particularly in winter. Previously, winter cold knocked out the beetles. With warmer conditions, the over-winter survival rate has jumped, allowing the beetles to complete their life cycle in one year rather than the customary two. The population has spiked. The larvae are filled with glycol, a natural antifreeze, which was not enough for the old forty below, but is now adequate to withstand milder cold. Now it's not cold enough to keep the beetle larvae in check over the winter. The beetles have other tricks, too. The immature beetles inject a fungus that prevents the tree from transporting sap. Without the fungus, the sap would drown the larvae. As it is, the trees quickly die. The beetle larvae hatch under the bark, where they consume the rich cambium layer that normally furnishes nutrients to the tree. It is mortal combat, and the trees typically lose. The standing, dead timber is ready to fuel the next runaway fire.

All told, the Rocky Mountains have witnessed the die-off of nearly seventy thousand square miles of forest—an area the size of Washington State—since 2000. Beetles, a deadly messenger of climate change, have turned huge swaths of forest from carbon sinks to carbon sources. Dead, decaying trees—especially when burning—add carbon dioxide to the atmosphere. If the climate continues to warm, the death toll and carbon loading will only magnify.

While Montana's outbreak of mountain bark beetles has not equaled the devastation of British Columbia and Colorado, the infestation is expanding and could reach three thousand square miles before long. Managers and scientists at Glacier National Park are watching the insect invasions to the north and south with trepidation.

"Our forests are in harm's way," says Fagre. "Right now, in the Park, there's a small outbreak around St. Mary's Lake. Nothing explosive yet, but in a warming world the bark beetles will be coming our way."

Meanwhile, other tipping points are looming right inside the Crown. The most obvious threshold event is the accelerating loss of glaciers. Entire watersheds are being degraded in their wake. Twenty-three drainages once had a glacier; today only fourteen. The co-opting of alpine tundra by an advancing tree line is another checkmate move, though on a longer timeline. Still, the mountain environment will change dramatically, losing perhaps ten percent or more of its alpine meadows.

Thus, in the Park, begins the march of homogeneity, the coming uniformity of the landscape. Fagre fears a mountain forest of wall-to-wall, dense pine trees that grow like weeds. Other habitats would evaporate. In this diminished world, the lynx—imprinted on one niche, one prey—could vanish like smoke.

Whether we call it an ecological threshold, environmental checkmate, or a tipping point, the effect on the ecosystem is similar: sudden, chaotic, and immense. For a biological community, the changing force is on the order of a tidal wave. It disrupts everything in sight. And like a wave, it spreads outward, the sphere of disturbance broadening until another ecosystem is touched.

The wave surges; the circle widens; the ripples fan out.

In his celebrated poem *The Second Coming*, William Butler Yeats likens this expanding gyre to the circling flight of a falcon—with a person at the center, looking skyward. Upon each rotation, the spiral extends. The bird wheels out of control like a kite at the end of a long string. Yeats laments, "The falcon cannot hear the falconer." The tether is stretched to the limit.

Yeats wrote his poem in 1920. There are many interpretations, ranging from naturalistic to societal. Much of the focus has been on the collapse of the traditional social order at the time. For example, younger generations were not mindful of their parents' values. The challenges of our century are on a larger canvas. The crisis today is the exhaustion of the world's resources. The relevance of the poem has shifted. We're in trouble now not because the "falcon cannot hear the falconer," but because we can no longer hear the falcon. We are so out of training, so out of tune, that even when nature calls out in distress, we don't hear the notes. There is a disconnect: The string has snapped.

Our deaf ear to nature has brought on a widening circle of

damage and neglect. We are seeing the fallout from this negligence at high latitudes and altitudes, the polar and alpine regions where glaciers reside. Globally, the most ubiquitous threshold response, other than within the oceans, is the melting of these ice fields. Like sentinels, the mountains and poles witness the onslaught first.

For this reason, alpine glaciers are more than just local indicators of climate change; their diminution constitutes a global signal announcing the ill health of the planet. The crisis begins at Glacier National Park but, as temperature climb elsewhere, the picture widens quickly to five continents around the globe:

European Alps: Largest and most celebrated of France's glaciers, the Mer de Glace ("sea of ice") is 6.8 miles (11 kilometers) long and 1,300 feet (396 meters) thick. It has thinned by twenty-seven percent, or 490 feet (150 meters) since 1907. For the entire Alps, total glacier volume is now one-third of the value estimated for 1850.

The Andes: Patagonian glaciers in Chile are retreating at a faster pace than in any other mountain range worldwide. The Northern Patagonian Ice Field, which has exit glaciers at its periphery, lost 36 square miles (93 square kilometers) of ice from 1946 to 1975 and 67 square miles (174 square kilometers) from 1975 to 1995. The rate is accelerating. Farther north, in Bolivia, the Chacaltaya Glacier vanished completely two years ago. Like Lima, La Paz is dependent on glacier melt for the city's water. In all, Andean glaciers have lost twenty percent of their volume since 1970.

Africa: The ice cover on the summit of Kilimanjaro has receded seventy-five percent since 1912. Scientists believe that drought—and thus the paucity of snow available to renew the ice—has dissipated the peak's glaciers more than elevated temperatures have.

When the glaciers disappear in 2033 or so, it will be the first time the summit has been barren in eleven thousand years.

Asia: The Gangotri Glacier, in the Garhwal Himalaya, has retreated 0.6 miles (1 kilometer) in thirty years. It supplies up to seventy percent of the water to the Ganges during the dry season, when monsoons don't fall. In turn, the Ganges supplies drinking water to 407 million people. All over the Himalayas and the Tibet-Qinghai Plateau, which together make up the largest body of ice outside the poles, glaciers are melting. (An exception is Pakistan's Karakoram Range, whose ice is advancing a bit.) In addition to the Ganges, six other great Asian rivers, from the Brahmaputra to the Mekong to the Yangtze, are fed by ice. If the glaciers disappear, the greatest famine in history will likely unfold.

North America: Besides the American Rockies and Cascades, major glaciers are positioned in Canada and Alaska. The Athabasca Glacier, which is an outlet to the huge Columbia Icefield in Alberta, has receded 4,900 feet (1.5 kilometers) since the late 1800s. In Denali National Park, all glaciers that have been surveyed are receding. Since 1976, when I was on an expedition to Mt. McKinley, the Muldrow Glacier—our route to Karsten's Ridge—has thinned over 66 feet (20 meters). A few Alaskan glaciers are surging, but most are pulling back into their cirques and valleys. Like other ice fields worldwide, their diminishment brings with it catastrophic impacts downstream.

How fast will the ice waste away? Computer simulations for the European Alps show that a 5.4 degree F (3 degree C) elevation of summer temperatures would reduce present glacier cover by eighty percent. A 9 degree F (5 degree C) increase would obliterate all the ice. The IPCC is predicting a rise of up to 7.2 degrees F (4 degrees C) by century's end, even if current trends of carbon emissions are cut significantly. So, the mountains—or, at least, the Alps—have perhaps a hundred years, unless we drastically change our habits. Just as important, Greenland and the West Antarctica Ice Sheet could melt significantly in that time. There's no certain timeline. But if all the ice on the polar caps were to ablate, the global ocean would rise an estimated 230 feet (70 meters).

Beyond their function as fountainheads for wildlife and humanity, glaciers have come to symbolize crystalline purity—what is natural and untainted. Genuine. And a little rogue. They often career down their slopes with violence. The quality of wildness is now perhaps best embodied in mountain glaciers and summits. The wild prairie has been partitioned; the rain forest is being cut; sand islands like Sable are being eroded; and wild rivers continue to be dammed. Even red cedar groves, our maritime forests, are disappearing. As the forests and glaciers decline, so does the cumulative wildness of the world.

Now, consider the roof of the planet: Alpine peaks have been relatively safe, untamable. They are remote and often inaccessible. They are, for the most part, inappropriate for development. This has been their salvation until now. They have remained wild.

The view from a mountain summit can grip the mind. When I close my eyes, I can still see the panorama from Swiftcurrent: nine shimmering glaciers in my view. Every vantage point in Glacier inspires reverence and awe. Perhaps more people should enjoy the alpine wilderness—to instill the broad respect it deserves. There's the conundrum: More visitors might heighten awareness but may also bring more wear and tear.

We see mountains as impregnable, as above the fray, but the

alpine world is as fragile as an ecosystem can be. While covering nearly one-quarter of the global land surface, mountains compress a wide profile of habitats into a relatively small area, thereby registering a high degree of biodiversity. Half of the U.N.'s thirty-four global biodiversity hot spots reside in mountain zones.

Yet the impact of footsteps along the world's backbone pales in comparison to the onslaught of a disrupted climate. When Sir Edmund Hillary climbed Everest in 1953, the Khumbu Glacier at its base was just beginning its modern decline. We have been whittling away at the ice ever since. (Khumbu now loses up to 65 feet (20 meters) per year.) When Dan Fagre visited Everest Base Camp on Khumbu in 2008, it was 130 feet (40 meters) lower in elevation, due to thinning of the glacier since Hillary's time. We look at Everest as the ultimate challenge. Now the greatest campaign may not be climbing but preserving its ice.

Reaffirming our emotional connection to glaciers and to all of nature is a good place to begin the campaign.

Wallace Stegner called wild country part of "the geography of hope." By this, he meant wilderness such as lofty mountains reassure us of our sanity as individuals. During the course of our lives, at some point or for the continuum, we exude a wild spirit. This wildness—the wind figuratively blowing through our hair—is much of what attracts us and connects us to wilderness, its rawness and purity. We long for its invitation to freedom.

Despite this attraction, we allow more wilderness to perish every year. Geographers estimate that between one-third and one-half of the Earth's land surface has been manipulated and transformed by human hands. Even the fields and forests of our youth, where we played and explored, have been subdivided. The fishing and swimming holes have dried up. And it's not just the acreage of wild places that has been diminished: The quality of the land has been impaired.

In the twentieth century, a national conservation movement sought to preserve what remained of wilderness. This preservation effort, culminating in the establishment of the national park system, is one of the great success stories of American democracy. Yet, with the borders of the parks and preserves now defined, we have become complacent. We assume the wilderness is safe: It has been set aside for posterity, after all. No one can touch it. And when extraction industries try to break the covenant, as in the battle over the Arctic National Wildlife Refuge, conservation groups keep the government honest. We count on them for protest. There is even an effort to foster wildlife migration corridors—for species like lynx and wolverine—between the refuges. This gives us comfort. But, while safeguarding the quantity of wilderness, we have forgotten to nurture the quality of wilderness, its character and health. In Glacier, as we have seen, the mountain acreage and mileage are preserved, but the glaciers are dying.

So we ask: How wild is our wilderness? How pure are Glacier National Park and the others?

The erosion of wildness began well before climate change became a household phrase. We have loved our parks and preserves into tameness. Some of our public lands—for example, Yellowstone—allow snowmobiles; some permit all-terrain vehicles (ATVs). The wildlife runs for cover. In Yosemite, car exhaust has harmed the trees. In Glacier, shuttle buses offer visitors a low-carbon alternative for exploring the Park. Nonetheless, more than fifty percent of the visitors stay in their cars and never go on a hike.

We have also studied and surveyed our parks into submission, placing our fingerprints on nearly everything. In Glacier, at least two mountain goats being fitted for radio collars died from the tranquilizer. The glaciers are not disturbed very much by monitoring, but, farther down the watershed, wildlife populations are disrupted. Do we really need to study grizzlies any more? Maybe they just need a million acres and to be left alone. (To their credit, grizzly researchers have recently abandoned radio collars in favor of profiling DNA of individual bears from hair left on scratching posts and trees.) Scientific invasiveness continues in some quarters. Would anyone cut down a stately cedar tree, just to examine the trunk rings, to age the species? It has happened before.

The results from all the investigations are invested in designing and managing the wilderness. Consequently, over the years, many parks have become too controlled to be truly wild. This speaks to process. Land or water is wild when it functions by its own schematic, unprompted by human hands. For over a century, we have tried to preserve things—acreage, habitats, species— when we actually should above all else preserve process, the workings of the ecosystem. The proper running of the machinery begins with the local climate and its watershed being in tune.

To this end, fisheries biologist Clint Muhlfeld says we should "think like a watershed." From the top of each watershed, namely the peaks and glaciers and snowfields, down through the creeks and canyons, past the streams and forests, and into the rivers and lakes, on down, the ecosystem has its governing principles and processes. We should respect them. The machinery of each watershed is fragile. "Aquatic systems are under assault," Clint says. "The whole hydrologic cycle is changing in the West."

Perhaps we should set aside as many watersheds as possible, then get out of their way. Let's follow the example of what the

Canadian Government has done for the single Atlantic drainage of Sable Island. They limit visitations and study. Our management and investigations can never be as wise as nature on her own.

Without watershed protection, or, on a larger scale, climateshed protection, the rivers warm up, silt in, run dry, and their basins burn over. In an evaporating climate space, cutthroat trout cede much of their old territory to rainbow trout. The native trout lose their genetic identity; streams merge with polluted waters. The hybrid cutthroat and the hybrid river are less fit. We have diluted the purity of the species and the habitat, as well.

Mountains and their glaciers, at the summit of the watersheds, are in the same predicament. For millions of years, a natural climate, with all its cycles and eccentricities, washed over the peaks. The plants and wildlife were adapted to the climate equilibrium and the flux of weather, within predictable bounds. Now, the climate is changing—drastically and suddenly. The resident species cannot adapt quickly. They must move or die. For many, migration isn't an option. As the native peaks hybridize with the civilized climate, pure wilderness is lost. Ecological hybridization leads to the extinction of authenticity. We have allowed our last pristine mountain wilderness to deteriorate in character for the bulk of the century.

But perhaps not forever. It could bounce back. We can change course by lowering the CO_2 in the atmosphere by 12.5 percent—from 400 parts per million to 350 parts per million, the acknowledged safe zone. This will require a much larger percentage decrease in carbon emissions on the ground; the exact figure is up for debate. Yet a recent article by British, German, and Swiss researchers in the journal *Nature* reports that a fifty percent reduction by 2050 may keep the temperature rise within 3.6 degrees F (2 degrees C), another threshold for safety. The prescription will not be cheap but it will save our glaciers, our coasts, and the heart of civilization. We'll know we're in the clear when European glaciers begin to recover in a generation or two after the cuts. The glaciers are the best gauge of our sanity. There are dozens of plans that prescribe how to do this—the possible ways to rebuild a healthy climate: clean coal, carbon taxes, renewable energy, and others. What is timely to explore here is why we should care to read these blueprints and act.

The urgency is there: glaciers melting, sea levels rising, marshland eroding, heat soaring, storms brewing, drought spreading, fires burning. Perhaps the loss of drinking and irrigation water is touching people the most. We are most dependent on water and food. In various parts of the world, water is falling into shorter and shorter supply. These are matters of self-interest, of survival. The mirror has two faces. For any community, either local or global, we must balance that self-interest with the common good. Fagre and Muhlfeld point to wilderness as one of those commonalities that urges us to address climate change. In mountain wilderness, we need to protect the climate space, not just to quench our thirst or to water our fields, but to allow the glaciers to proceed with all their ecological gifts and value. Alpine wilderness, its quality and purity and process, is one more reason we should avoid overtaxing the planet.

We know about the growing population of climate refugees. Now, thanks to Fagre and his colleagues, we know a little more about the starving streams above them, the naked rock where glaciers once flowed.

And yet, why should we care personally, viscerally, about the integrity of alpine wilderness, even if it is our common ground?

Why should we be concerned about something so distant, maybe a couple of days' drive away?

Perhaps it's closer than we think.

The wild spirit in all of us identifies with nature. We long for it. After all, humanity sprang forth from the natural world. If that spirit has ebbed in recent years, there are ways to reinvigorate it. We can go for a hike in Glacier or another park. Take the grandparents. Take the children. Here is our motivation: The most recent generation—that comprising youth under thirty—has been labeled as having "nature-deficit disorder." These young people are more accustomed to technology than the outdoors. With little emotional connection to nature, so the charge goes, this "tech generation" is destined to be a poor steward of the Earth. There are countless statistics to back this up. Visitations to national parks dropped nearly twenty-five percent between 1987 and 2003. Non-environmentalists are less likely to have had childhood nature experiences than staff members of conservation organizations. The ranks of government environmental agencies are the same. Look at the early years of Fagre and Muhlfeld, of Lisa and Lindsey, of Kevin and Erich—all nature kids. Baby boomers—those born between 1946 and 1964—may be the last American generation to hold an intimate attachment to the outdoors.

A solution is at hand: Initiation into nature—into wildness—could make younger people care. Walking in a cedar grove, like the one I'm in right now, or watching a grizzly—you just can't help it. You feel exhilarated. Wonder is visceral; it is infectious. Wonder breeds reverence.

There's another question. Even if we care—complete with awe and humility—why should we act? Why should we inconvenience ourselves with protesting or preventing climate change? Our vote, our voice, our action may be needed: Even if we eventually get out of wilderness's way, we will have to fix the atmosphere first. We need to reverse the damage to the climate. But why should any one of us step up and make that stand?

"It's a moral issue, not a scientific one," says Dan Fagre, cutting to the heart of the matter. "Science is like a navigator on a ship. It can point the way. It can enlighten people on which direction to take. That's all. The crew has to do the heavy lifting, out of some sense of responsibility. If we have enlightenment, the way forward becomes obvious."

Ethicists enlist the story of a burning building to illustrate the moral imperative. If we are the first to spot a burning building or forest or ship, and we know lives are endangered, we consider it our moral obligation to help the victims, or at least report the fire. The moral necessity revolves around the plight of human suffering—our ability to help and our proximity to the crisis.

All ethics rests upon the premise that each of us is a member of a community. We are torn between self-interest and altruism. Sometimes we walk by the fire; other times, we stop to help. All the while, we know what's right. Morality calls on us to assist in each communal crisis. Aldo Leopold expanded the borders of community to encompass land, plants, animals, and ecosystems. There may be many threatened by fire in a watershed. Or threatened by global warming. We have a moral obligation to act—to arrest climate change—because people are suffering and, if you embrace the ecological argument, because wilderness is suffering as well. On top of this, it's in our self-interest to have access to water. Through our use of energy, in our everyday choices, the health of the atmosphere no longer remains a remote issue or someone else's problem. We can choose to drive a car or ride a bike. It manifests for each of us, hour by hour, all day long.

"To halt climate change," says Fagre, "we will have to rely on the better angels of human nature. The challenge is unique in human history—a global imperative. Ethical action will be required of governments and individuals, together and at once."

That action might center around some defining ecological principles. Leopold again: "A thing is right when it tends to preserve the integrity, stability, and beauty of the biotic community. It is wrong when it tends otherwise." To this "land ethic" we might add a watershed addendum: "Stay out of nature's way by giving wide berth to the watershed, its processes and systems." Every drainage is an ecosystem from glaciers and headwaters to the sea. We must maintain its purity. For starters, we need to unclog the atmosphere and restore the natural climate. Otherwise, man's intrusion will result in more hybridization of the wilderness. We have no wild country left to spare.

Standing in the cedar grove, these are my thoughts. I can still hear the thrush's call. But, like Dan Fagre, I am caught between two worlds: the geology of the past and the ecology of the future. The local, ancient glaciers melt before my eyes, but the world has promise of redemption. Like Fagre, I believe in science but in humanity, too. I am also torn on how best to preserve my memories: whether to paint in my mind the way it once was—spectacular ramps of blue ice—or how it has become. I have memories of vanished wilderness that haunt me like a shadow. Mountains of dreams. For this lost provenance, I mourn. For the wild places remaining, I reap some solace that they may survive.

I can close my eyes and still see the blue-and-white shimmering faces of Grinnell, Sperry, and Swiftcurrent. I can hear the ravens chattering, as they lift gently from the snows. Montana's ice fields may be vanishing but, in their last days, they are not silent. The glaciers are creaking and groaning, and trickling with pure water. The mountains are speaking out. Listen to the raven. Listen to the falcon.

Like Fagre, listen to the ice.

Selected Bibliography

Adams, Jonathan. "Rising Sea Levels Threaten Small Pacific Island Nations." *The New York Times* (May 3, 2007): 1–2.

Allendorf, Fred W., R. F. Leary, N. P. Hitt, K. L. Knudsen, M. C. Boyer, and P. Spruell. "Cutthroat Trout Hybridization and the U.S. Endangered Species Act: One Species, Two Policies." *Conservation Biology* (2005): 1326–1328.

Allendorf, Fred W., Robb F. Leary, Paul Spruell, and John K. Wenburg. "The Problem with Hybrids: Setting Conservation Guidelines." *Trends in Ecology and Evolution* 16 (2001): 613–622.

Armstrong, Betsy R., and Knox Williams. *The Avalanche Book*. Golden, Colorado: Fulcrum Publishing, 1992.

Baker, William L. *Fire Ecology in Rocky Mountain Landscapes*. Washington, D.C.: Island Press, 2009.

Balog, James, and the Extreme Ice Survey. *Ice: Portraits of Vanishing Glaciers*. New York: Rizzoli, 2012.

Bear, Elizabeth A., Thomas E. McMahon, and Alexander V. Zale. "Comparative Thermal Requirements of Westslope Cutthroat Trout and Rainbow Trout: Implications for Species Interactions and Development of Thermal Protection Standards." *Trans. Am. Fisheries Society* 136 (2007): 1113–1121.

Beever, E. A., P. F. Brussard, and J. Berger. "Patterns of Extirpation Among Isolated Populations of Pikas *(Ochotona princeps)* in the Great Basin." *J. of Mammalogy* 84 (2003): 37–54.

Beever, E. A., and J. L. Wilkening. "Playing by New Rules: Altered Climates Are Affecting Some Pikas Dramatically—and Rapidly." *The Wildlife Professional* 5 (2011): 38–41.

Blond, Becca. "As Politicians Waffle on Climate Change, Glaciers Exit Glacier National Park." *The Colorado Independent* (February 12, 2010): 1–2.

Buccholtz, C. W. *Man in Glacier.* West Glacier, Montana: Glacier National History Association, 1976.

Butler, David R., George P. Malanson, Stephen J. Walsh, and Daniel B. Fagre. *The Changing Alpine Treeline: The Example of Glacier National Park, MT, USA.* Developments in Earth Surface Processes, 12. Amsterdam: Elsevier, 2009.

Carrara, P. E. *Late Quaternary Glacial and Vegetative History of the Glacier National Park Region, Montana.* Denver: United States Government Printing Office, 1989.

Carrara, P. E., and R. G. McGimsey. "Distribution of Moraines and Extent of Glaciers from Mid-19th Century to 1979, Glacier National Park, Montana." *U.S. Geological Survey* (1988): Map I-1508-C.

Chadwick, Douglas H. *A Beast the Color of Winter: The Mountain Goat Observed.* Lincoln, Nebraska: Bison Books, 2002.

Chadwick, Douglas H. *The Wolverine Way.* Ventura, California: Patagonia Books, 2010.

Chaudhary, Sunita. "Save Our Mountains: Why Mountain Diversity Matters." *The Kathmandu Post* (May 13, 2011): 1–2.

Copeland, J. P., K. S. McKelvey, K. B. Aubry, A. Landa, J. Persson, R. M. Inman, J. Krebs, E. Lofroth, H. Golden, J. R. Squires, A. Magoun, M. K. Schwartz, J. Wilmot, C. L. Copeland, R. E. Yates, I. Kojola, and R. May. "The Bioclimatic Envelope of the Wolverine *(Gulo gulo)*: Do Climatic Constraints Limit Its Geographic Distribution?" *Canadian J. Zoology* 88 (2010): 233–246.

Crutchfield, James A. *It Happened in Montana*. Guilford, Connecticut: Globe Pequot Press, 2008.

DiMento, Joseph F. C., and Pamela Doughman (eds.). *Climate Change: What It Means for Us, Our Children, and Our Grandchildren*. Cambridge, Massachusetts: MIT Press, 2007.

Edwards, J. Gordon. *A Climber's Guide to Glacier National Park*. Helena, Montana: Falcon Press Publishing, 1995.

Ehrlich, Gretel. *The Future of Ice: A Journey into Cold*. New York: Vintage, 2005.

Erschbamer, Brigitta, Thomas Kiebacher, Martin Mallaun, and Peter Unterluggauer. "Short-Term Signals of Climate Change Along an Altitudinal Gradient in South Alps." *Plant Ecology* 202 (2009): 79–89.

ETH Zurich. "Climate Change: Halving Carbon Dioxide Emissions by 2050 Could Stabilize Global Warming." *Science Daily* (May 4, 2009).

Extreme Ice Survey. "What Is the Extreme Ice Survey?" Accessed June 16, 2012. http://www.extremeicesurvey.org/index.php/about/.

Fagre, Daniel. "Changing Mountain Landscapes in a Changing Climate: Looking into the Future." *Changing Landscapes* (Summer 2000): 31–39.

Fagre, D. B., P. L. Comanor, J. D. White, F. R. Hauer, and S. W. Running. "Watershed Responses to Climate Change at Glacier National Park." *Journal of Am. Water Resources Assoc.* 33 (1997): 755–765.

Fagre, Daniel B., and Erich H. Peitzsch. "Avalanche Ecology and Large Magnitude Avalanche Events—Glacier National Park, Montana, USA." *U.S. Geological Survey* (2010): 1–6.

Fagre, Daniel B., David L. Peterson, and Amy E. Hessl. "Taking the Pulse of Mountains: Ecosystem Responses to Climatic Variability." *Climatic Change* 59 (2003): 263–282.

Ferguson, Sue A. *Glaciers of North America: A Field Guide*. Golden, Colorado: Fulcrum Publishing, 1992.

Garvey, James. *The Ethics of Climate Change: Right and Wrong in a Warming World*. London: Continuum International Publishing Group, 2008.

Geranios, Nicholas K. "Glacier at 100: Age Has Not Been Kind." *USA Today* (May 11, 2010): 1.

Gildart, R. C. *Glacier Country: Montana's Glacier National Park*. Helena, Montana: American Geographic Publishing, 1990.

GlacierWorks. "The Mission of GlacierWorks." Accessed May 5, 2013. http://www.glacierworks.org.

Global Observation Research Initiative in Alpine Environments (GLORIA). "The Purpose of GLORIA." Accessed June 17, 2012. http://www.gloria.ac.at/?a=2.

Gordon, John. *Glaciers*. Stillwater, Minnesota: Voyageur Press, 2001.

Gosnell, Mariana. *Ice: The Nature, the History, and the Uses of an Astonishing Substance*. Chicago: University of Chicago Press, 2005.

Guthrie, C. W. *Glacier National Park: The First 100 Years*. Helena, Montana: Farcountry Press, 2008.

Hall, Myrna H. P., and Daniel B. Fagre. "Modeled Climate-Induced Glacier Change in Glacier National Park, 1850–2100." *Bioscience* 53 (2003): 131–140.

Hart, Jeff. *Montana Native Plants and Early Peoples*. Helena, Montana: Montana Historical Society Press, 1996.

Huang, Lily. "The Case of the Disappearing Rabbit: Ten Million Acres of the American West May Depend on the Fate of the Vanishing Snowshoe Hare." *Newsweek* (July 25, 2009): 11–14.

Selected Bibliography

Jenkins, McKay. *The White Death: Tragedy and Heroism in an Avalanche Zone.* New York: Anchor Books, 2001.

Joint Fire Science Program. "The Fire-Climate Connection." *Fire Science Digest* 1 (2007): 2–11.

Kimball, Shannon F., and Peter Lesica. *Wildflowers of Glacier National Park and Surrounding Areas.* Kalispell, Montana: Trillium Press, 2005.

Klasner, Frederick L., and Daniel B. Fagre. "A Half Century of Change in Alpine Treeline Patterns at Glacier National Park, Montana, U.S.A." *Arctic, Antarctic, and Alpine Research* 34 (2002): 49–56.

Kolbert, Elizabeth. *Field Notes from a Catastrophe: Man, Nature, and Climate Change.* New York: Bloomsberg, 2006.

Kricher, John C., and Gordon Morrison. *Ecology of Western Forests.* Peterson Field Guide Series. Boston: Houghton Mifflin Company, 1993.

Lanza, Michael. "Ghosts of Glacier: The Park's Rivers of Ice Have Flowed for 7,000 Years; They'll Be Gone in 30." *Backpacker Magazine* (September 2007): 23–29.

Leopold, Aldo. *A Sand County Almanac: And Sketches Here and There.* New York: Oxford University Press, 1949.

Lesica, P., and B. McCune. "Decline of Arctic–Alpine Plants at the Southern Margin of Their Range Following a Decade of Climatic Warming." *Journal of Vegetation Science* 15 (2004): 679–690.

Lopez, Barry (ed.). *The Future of Nature: Writing on a Human Ecology from Orion Magazine.* Minneapolis, Minnesota: Milkweed Editions, 2007.

Louv, Richard. *Last Child in the Woods: Saving Our Children from Nature-Deficit Disorder.* Chapel Hill, North Carolina: Algonquin Books, 2008.

Louv, Richard. *The Nature Principle: Human Restoration and the End of Nature-Deficit Disorder.* Chapel Hill, North Carolina: Algonquin Books, 2011.

Lovejoy, Thomas E., and Lee Hannah (eds.). *Climate Change and Biodiversity*. New Haven: Yale University Press, 2005.

Macdougall, Doug. *Frozen Earth: The Once and Future Story of Ice Ages*. Berkeley, California: University of California Press, 2004.

Macfarlane, Robert. *The Wild Places*. New York: Penguin Books, 2007.

Mann, Michael E., and Lee R. Kump. *Dire Predictions: Understanding Global Warming*. New York: Dorling Kindersley, 2008.

Mathews, Daniel. *Rocky Mountain Natural History: Grand Teton to Jasper—A Trailside Reference*. Portland, Oregon: Raven Editions, 2003.

McClung, David, and Peter Schaerer. *The Avalanche Handbook*. Seattle, Washington: Mountaineer Books, 2006.

McKibben, Bill. *The End of Nature*. New York: Random House, 1989.

McKibben, Bill. *Eaarth: Making a Life on a Tough New Planet*. New York: St. Martin's Press, 2011.

Millar, Connie, and Daniel Fagre. "Monitoring Alpine Plants for Climate Change: The North American GLORIA Project." *Mountain Views: The Newsletter of the Consortium for Integrated Climate Research in Western Mountains (CIRMOUNT)* 1 (2007): 12–14.

Miller, Char (ed.). *Water in the 21st-Century West: A High Country News Reader*. Corvallis, Oregon: Oregon State University Press, 2009.

Moore, Kathleen Dean, and Michael P. Nelson (eds.). *Moral Ground: Ethical Action for a Planet in Peril*. San Antonio, Texas: Trinity University Press, 2010.

Moravek, Vince. *It Happened in Glacier National Park*. Guilford, Connecticut: Globe Pequot Press, 2005.

Muhlfeld, Clint C., Joseph Giersch, F. Richard Hauer, Gregory T. Pederson, Gordon Luikart, Douglas P. Peterson, Christopher C. Downs, and Daniel B. Fagre. "Climate Change Links Fate of Glaciers and an Endemic Alpine Invertebrate." *Climatic Change* 106 (2011): 337–345.

Muhlfeld, Clint C., Steven T. Kalinowski, Thomas E. McMahon, Mark L. Taper, Sally Painter, Robb F. Leary, and Fred W. Allendorf. "Hybridization Rapidly Reduces Fitness of a Native Trout in the Wild." *Biology Letters* 5 (2009): 328–331.

Muhlfeld, Clint C., T. E. McMahon, M. C. Boyer, and R. E. Gresswell. "Local Habitat, Watershed, and Biotic Factors Influencing the Spread of Hybridization between Native Westslope Cutthroat Trout and Introduced Rainbow Trout." *Trans. Am. Fisheries Society* 138 (2009): 1036–1051.

Nash, J. Madeleine. *El Niño: Unlocking the Secrets of the Master Weather-Maker.* New York: Warner Books, 2002.

North Cascades Glacier Climate Project. "Glacier Mass Balance Forecasting." Accessed June 16, 2012. http://www.nichols.edu/depart ments/glacier/mass%20balance%20forecasting.htm.

Oelschlaeger, Max. *The Idea of Wilderness: From Prehistory to the Age of Ecology.* New Haven: Yale University Press, 1991.

O'Hanlon, Larry. "Sea Level Rise May Sink Island Nations." *Discovery News* (February 3, 2010): 1–2.

Olsen, Jack. *Night of the Grizzlies.* Moose, Wyoming: Homestead Publishing, 1996.

Pederson, Gregory T., Lisa J. Graumlich, Daniel B. Fagre, Todd Kipfer, and Clint C. Muhlfeld. "A Century of Climate and Ecosystem Change in Western Montana: What Do Temperature Trends Portend?" *Climatic Change* 98 (2010): 133–154.

Pederson, Gregory T., Stephen T. Gray, Toby Ault, Wendy Marsh, Daniel B. Fagre, Andrew G. Bunn, Connie A. Woodhouse, and Lisa J. Graumlich. "Climatic Controls on the Snowmelt Hydrology of the Northern Rocky Mountains, USA." *Journal of Climate* (2011): 1–22.

Peters, Debra P. C., Jin Yao, and Kris M. Havstad. "Insights to Invasive Species Dynamics from Desertification Studies." *Weed Technology* 18 (2004): 1221–1225.

Phillips, H. Wayne. *Northern Rocky Mountain Wildflowers*. Helena, Montana: Falcon Guides, 2001.

Pilkey, Orrin H., and Rob Young. *The Rising Sea*. Washington, D.C.: Island Press, 2009.

Raup, Omar B., Robert L. Earhart, James W. Whipple, and Paul E. Carrara. *Geology: Along Going-to-the-Sun Road, Glacier National Park, Montana*. West Glacier, Montana: Glacier Natural History Association, 1983.

Reardon, Blase A., and Daniel B. Fagre. "Reassessing a Troublesome Fact of Mountain Life: Avalanches in Glacier National Park." *Park Science* 24 (2006): 1–3.

Reardon, B. A., J. T. Harper, and D. B. Fagre. "Mass Balance of a Cirque Glacier in the U.S. Rocky Mountains." Proceedings of Mass Balance Measurement and Modeling Workshop, Skeikampen, Norway. (2008): 1–5.

Reardon, B. A., G. T. Pederson, G. J. Caruso, and D. B. Fagre. "Spatial Reconstructions and Comparisons of Historic Snow Avalanche Frequency and Extent Using Tree Rings in Glacier National Park, Montana, U.S.A." *Arctic, Antarctic, and Alpine Research* 40 (2008): 148–160.

Reis, Patrick. "Climate Concerns Prompt Lawsuit over Lynx Habitat." *The New York Times/Greenwire* (May 27, 2009): 1–2.

Reisner, Marc. *Cadillac Desert: The American West and Its Disappearing Water*. New York: Penguin Books, 1993.

Robbins, Jim. "Spread of Bark Beetles Kills Millions of Acres of Trees in West." *The New York Times* (November 18, 2008): D3.

Rockwell, David. *Glacier: A Natural History Guide*. Helena, Montana: Falcon Guides, 2007.

Rosner, Hillary. "The Color of Bunny: Can Evolution Save Species from Climate Change?" *High Country News* (February 6, 2012): 11–16.

Schneider, Russ. *Fishing Glacier National Park*. Helena, Montana: Falcon Guides, 2002.

Selters, Andy. *Glacier Travel and Crevasse Rescue*, 2nd ed. Seattle, Washington: The Mountaineers, 1999.

Streever, Bill. *Cold: Adventures in the World's Frozen Places*. New York: Little, Brown and Company, 2009.

Strickler, Dee. *Alpine Wildflowers: Showy Wildflowers of the Alpine and Subalpine Areas of the Northern Rocky Mountain States*. Columbia Falls, Montana: The Flower Press, 1990.

Turner, Jack. *The Abstract Wild*. Tucson, Arizona: The University of Arizona Press, 1996.

Thybony, Scott. *Wildfire*. Tucson, Arizona: Western National Parks Association, 2002.

Ulrich, Tom J. *Mammals of the Northern Rockies*. Missoula, Montana: Mountain Press Publishing, 1986.

United Nations Intergovernmental Panel on Climate Change (IPCC). "IPCC Fourth Assessment Report, 2007." Accessed June 22, 2012. http://www.ipcc.ch/publications_and_data/publications_and_data _reports.shtml.

University of Arizona. "More Large Forest Fires Linked to Climate Change." *Science Daily* (July 10, 2006): 1–3.

U.S. Climate Change Science Program. *Thresholds of Climate Change in Ecosystems*. Reston, Virginia: U.S. Geological Survey, 2009.

U.S. Global Change Research Program (USGCRP). "Global Climate Change Impacts in the United States." Washington, D.C.: USGCRP, 2009. Accessed December 12, 2012. http://www.globalchange.gov/usimpacts.

USGS, Glacier & Snow Program of Alaska and Washington Science Centers. "Benchmark Glaciers." Accessed June 16, 2012. http://ak.water .usgs.gov/glaciology/index.html.

USGS, Northern Rocky Mountain Science Center (NOROCK). "Gigapixel Imagery of Glacier National Park." Accessed June 15, 2012. http://www.nrmsc.usgs.gov/GNP_gigapan.

USGS, Northern Rocky Mountain Science Center (NOROCK). "Glacier Research: Climate Change in Mountain Ecosystems (CCME)." Accessed June 15, 2012. http://www.nrmsc.usgs.gov/research/glacier _research.htm.

USGS, Northern Rocky Mountain Science Center (NOROCK). "Modeled Climate-Induced Glacier Change in Glacier National Park, 1850–2100." Accessed June 15, 2012. http://www.nrmsc.usgs.gov/research /glacier_model.htm.

USGS, Northern Rocky Mountain Science Center (NOROCK). "Repeat Photography Project." Accessed June 15, 2012. http://www.nrmsc .usgs.gov/repeatphoto/overview.htm.

Walter, Dave. *Montana Campfire Tales: Fourteen Historical Narratives*. Guilford, Connecticut: Globe Pequot Press, 1997.

Westerling, A. L., H. G. Hidalgo, D. R. Cayan, and T. W. Swetnam. "Warming and Earlier Spring Increases Western U.S. Forest Wildfire Activity." *Science Express* (July 6, 2006): 1–5.

Whitney, Stephen. *Western Forests*. New York: Knopf, 1985.

Wiley–Blackwell. "American Pikas: Contemporary Climate Change Alters the Pace and Drivers of Extinction." *Science Daily* (April 21, 2011): 1–2.

Woody, Todd. "U.S. Fish and Wildlife Service Says Pika Not Imperiled by Climate Change." *The New York Times/Green Blog* (February 8, 2010): 1–2.

World Glacier Monitoring Service (WGMS). "Welcome to the World Glacier Monitoring Service." Accessed June 18, 2012. http://www.geo .uzh.ch/microsite/wgms/.

350.org. "350 Science." Accessed June 18, 2012. http://www.350.org/en/ about/science.

Acknowledgments

Serendipity plays a role in every investigation and every adventure. This axiom rang true on my journey through Glacier National Park, where each bend in the trail offered up a surprise. Some of these revelations were troubling, namely the deterioration of watersheds from glacier to stream to valley floor. But others were welcome. I had a run of good luck in meeting the spirited staff of the U.S. Geological Survey, high on my list of enlightened, chance encounters. I was fortunate to cross paths with Dan Fagre and Clint Muhlfeld of the USGS, the former on glaciers high in the mountains and the latter along the alpine streams fed by ice and snow. Both scientists shared their research and personal histories with me, theorized about the future, and introduced me to other professionals in the field. Over the four years of the project, I always found them engaging and ready to help. I am indebted to both of them, as the guiding lights of this story.

My appreciation also goes to the USGS staff who work tirelessly on the ice to bring home the vital data on glaciers and associated wilderness. Among the crew were Lisa McKeon, Lindsey Bengtson, Kevin Jacks, Erich Peitzsch, and Chris Miller. All granted my request for an interview. I thank them for their patience and good humor in the face of my endless queries.

Other scientists and staff in West Glacier, mostly with the National Park Service, were helpful in a variety of ways. I would like to thank Mark Biel (NPS), Jami Belt (NPS), Tara Carolin (NPS), Joe Giersch (USGS), Lon Johnson (NPS), John Waller (NPS), and Dennis Divoky (NPS), who kindly took me on a tour of the Robert Fire. Christine Baker also offered assistance. Summer rangers, including Becky, Teagan, and Bill, helped me out on the trail by explaining geology and identifying plants and wildlife.

Around the periphery of the Park, I enlisted the help of Stacey Bengtson, Mike Davies, and Judith Moore of Glacier Guides, Inc., who led the way to Sperry Chalet, Granite Park Chalet, and the Garden Wall. Stephan Prince of the Glacier Natural History Association directed me to several books that proved invaluable. On the east side, Ed Des Rosier and Keith Schauf directed me to unexpected vistas.

Outside of the Park (and, in several cases, outside Montana), many scientists sent articles or offered their expertise, including Kurt Fausch, Robert Keane, Peter Lesica, Rachel Loehman, George Malanson, Orrin Pilkey, Jason Sibold, and James Wang. The Environmental Defense Fund, 350.org, and the IPCC were valuable sources of general information on climate change.

At home in Santa Fe, Susanna Hesch, my researcher and typist, made sure each draft arrived on schedule. She also organized and verified the bibliography. I thank her for her friendship and all the care and dedication she brought to the project.

My first reader of every draft and my most creative critic has been my partner, Donna Grosvenor. She read the book countless times and made many important suggestions. While a good eye is helpful, the best editorial advantage comes from a great ear. Donna has a fine ear, a gift for the language.

I am grateful for my friends and colleagues who read chapters or offered encouragement along the way: Pam Adger, Susan Armenti, Dan and Tammy Beigel, Walter and Diane Burke, Andrea Campbell, Rick Carey, Marilyn Colborn, Jeff Davis, Frank Fitzgerald, Bill Fritz, Linda Fritz, Roger Hart, Chris Haws, Laura Helmuth, Melanie Hornstein, Jake Kritzer, Will and Eleanor Lockwood, Brother Doug Lyman, Jim Lynn, Elizabeth Lyon, Dorothy Massey, Bill and Kay McConnell, Marilyn McDonald, Aren Minter, Melinda Morrison, Gretchen Paffendorf, Jack and Rebecca Parsons, George Reiger, Claire Romero, Rebecca Schwarzlose, Dorothy Sherwood, Bill Sidman, Jake Stout, and our Maine coon cats, Tasa and Dini.

Early in the project, my agent, Lauren MacLeod, and Wendy Strothman, both at the Strothman Agency, critiqued the first chapter and made many helpful suggestions that set the tone for the rest of the book. I thank them for their encouragement along the way. Michael Flamini, my editor at St. Martin's Press, was enthusiastic about the story from the beginning. He suggested the introduction that is included here. I thank him for his editorial guidance. My gratitude also goes out to Vicki Lame at St. Martin's for her care and indispensability.

I also benefited from the support and humor of my children, Hovey and Lexi, and their spouses, Kristina and David. A conversation with David about Edward Abbey gave me an insight in how to approach the epilogue. And, finally, I cannot find the words to express the debt I owe Donna. She kept the fire burning while I was writing and strategically fanned the flames so we could see where the story led. The personal narratives, mostly about my mountaineering adventures, included here, were her idea. Those expeditions and encounters with nature prepared me for Glacier. But the most serendipitous moment in those journeys is simply that the trail of adventure brought me to her door. It's always thrilling to come home again, across that threshold, every day fresh after all these years.